European Communities Oil and Gas Research and Development Projects

First Status Report

European Communities Oil and Gas Research and Development Projects

First Status Report

Compiled by

Derek Fee

Commission of the European Communities,
Directorate-General for Energy,
Brussels

Published by
Graham & Trotman
for the Commission of the European Communities

Published in 1982 by
Graham & Trotman Limited
Sterling House, 66 Wilton Road
London SW1V 1DE, United Kingdom

for the Commission of the European Communities, Directorate-General
Information Market and Innovation

EUR 8043

ISBN-13:978-94-009-7373-2   e-ISBN-13:978-94-009-7371-8
DOI:10.1007/978-94-009-7371-8

Legal Notice

CONTENTS

P R E F A C E

The Community's Hydrocarbon Project Scheme was the first concrete
response of the Community to the energy supply crisis of 1973.
The purpose of the programme is to encourage the development of
oil and gas supplies by subsidising technological development
projects in the hydrocarbon sector.

The programme has now been running 8 years and many significant
advances have been made.

Technological progress will largely condition the level of oil and
gas supplies available to both industrialised and developing nations.

The purpose of this report is to publicise what has been achieved
in the fields of exploration, production, transportation and
storage technologies during the first six rounds of the programme.

The report bears witness to the dynamism of European enterprises,
and to their readiness and ability to innovate and to create the
very advanced technologies essential to the development of
hydrocarbon resources.

C.J. AUDLAND
Director-General

Directorate-General for Energy.

INTRODUCTION

The Commission of the European Community has, by means of the Directorate General for Energy, been involved in energy research aimed at improving the energy supply situation of the Community. This involvement is on two levels, firstly the Community supports research and development aimed at improving the technologies associated with the location and production of traditional fuels and secondly the Community is actively involved in research to replace traditional energy sources with suitable alternatives.

Given the parlous state of the energy supply situation in the Community, it was felt that a special effort was required to develop new technologies associated with improving the supply of traditional fuels and in developing and establishing alternative sources of energy.

The initiative of the Community was begun in 1973 when the Council approved Regulation (EEC) 3056/73 setting up a series of three year research and development programmes in the oil and gas sector. This programme was one factor in the Community's response to the supply crisis of 1973.

The purpose of this report is to present the research and development carried out under contract in the framework of this programme, which is aimed at the improvement of the oil and gas supply situation by developing these technologies associated with the location, production and transportation of oil and gas. The report begins by examining the oil and gas programme in the context of energy research and development within the Directorate General for Energy. It then presents a summary of the programme's content, its implementation and supervision structures and the results to date. Finally, a detailed summary of each of the contracts concluded up to the sixth round of projects (1980). This summary which constitutes the main part of this report is presented on a per round basis and not by subject area.

COMMUNITY ENERGY RESEARCH AND DEVELOPMENT STRATEGY

The energy strategy of the Community has been developed in the context of the fact that 60% of the energy supplies of the Community are imported. The strategy has therefore evolved on three levels ; firstly by endeavouring to secure traditional energy supplies for the Community, secondly by embarking on an energy saving programme aimed at reducing the demand for energy within the Community without at the same time causing economic disruption and thirdly by developing alternative energy sources which may in time supplant traditional forms of energy.

These considerations also influence the development of the Community's energy research and development programmes, which in the case of the Directorate General for Energy can be divided into four strategic sectors:

1) The improvement of supply of conventional energy sources by research and development in hydrocarbon technology.

2) The promotion of technological advance in the exploitation of alternative energy sources.

3) The identification and demonstration of technologies in the field of energy saving.

4) The improvement, by an active exploration programme, of the Community's supply of uranium.

There are, of course, variable lead times in the implementation of the various technologies developed in the programmes supported by the Community. While many of the technologies developed under the hydrocarbons programme have already become state of art many of the developments in the energy saving and alternative energy sources sector have not yet been fully implemented. However, given the long lead times associated with these programmes one should not underestimate the strategic importance to the Community of these two vital sources.

The hydrocarbon technological development programme has been concentrated on developing techniques which should improve the supply situation of the Community in the shortest possible time. A good example of this type of technology is enhanced recovery, the development of which is already having a considerable impact on the ultimate recoverable reserved of crude oil, both within the Community and world wide.

THE COUNCIL PROGRAMME DECISION

The decision of the Council adopting Regulation (EEC) 3056/73 was taken on the 9th of November 1973. This Regulation set up a series of three year research and development projects in the hydrocarbon sector.

Article 1. of the Regulation aptly describes the purpose of the programme,

"The Community may in accordance with the conditions laid down hereinafter, grant financial support, in so far as this is essential, for the carrying out of projects (Community projects) which are of fundamental importance in ensuring the Community's supply of hydrocarbons".

The projects carried out thus far have been subdivided into the following fields:

- exploration including seismology
- drilling
- production
- enhanced recovery
- auxiliary ship, submersibles and navigation systems
- pipelaying
- transportation by pipeline
- transportation by ship
- gas technology
- storage

The amount of support allocated is generally expressed in accordance with the chances of success of the projects and their relative importance for the Community. Projects that, by exploitation of the results, will mean an increase in resource and/or an acceleration in the exploration and exploitation of hydrocarbon resources in the Community, have the benefit of the maximum rate of support of 40 to 45 percent. Transport and storage projects usually receive medium support of 30 to 35 percent. Projects that concern the supply of services have minimum support of about 25 percent.

Table 1 shows the sectorial distribution and the global funding situation for the first six rounds of projects.

IMPLEMENTATION AND SUPERVISION STRUCTURES

The Commission has several responsabilities in the implementation of Regulation (EEC) 3056/73. Firstly the Commission is responsible for issuing the call for tenders and for examining the proposals submitted. A proposal is then drawn up and submitted for discussion within the Energy Working Group of the Council. When a Council decision is taken the services of the Commission are then responsible for the negociation and conclusion of the contracts and the monitoring of the project, both technically and financially during its lifetime. The technical co-ordination of the programme is the responsibility of Directorate C "Hydrocarbons" while the Contracts Division of Directorate A is responsible for all administrative matters.

After the signature of the contract the projects are monitored both by examination of the periodic reports and by making annual technical and financial controls on site.

Annual reports concerning the state of advancement of the programme are made to the European Parliament and the Council.

Since the requirements for research and development in the oil and gas sector are constantly changing the criteria for examining proposals are constantly reviewed. When a major change in criteria is proposed a review meeting is held with the various national experts associated with the Energie Working Group of the Council.

STATUS OF IMPLEMENTATION AND FIRST RESULTS

A subsidy has been granted to 180 projects covering rounds 1 to 6 of the programme. Many of the projects have been completed but as can been seen from the summaries many are still in progress. It has been our experience on this programme that while a small proportion of projects have been completed before or on schedule the great majority have suffered delays which caused them to overrun the projected finishing date.

In terms of technical results, the programme has already contributed to the solution of many technical problems which were considered by experts in the field as bottlenecks.

While the summaries contain details of the results of individual projects the following sections give some indication of the global situation.

- <u>1975 - 1st Round Projects</u>

The most significant results were obtained in the areas of pipelaying, in deep sea drilling and in deep water production techniques.

The <u>pipelaying</u> tests in the Straits of Messina and the Sicily Channel carried out by SNAM were directly responsible for proving the viability of deep water pipelaying in the Mediterranean and have led to the construction of the gas line from Algeria to Italy which will carry 12 billion m^3 of natural gas per year.

In <u>deep sea drilling</u>, several new pieces of equipment, especially the drilling equipment of the drillship "Petrel", were developed by GERTH (Groupement Européen de Recherches Technologiques sur les Hydrocarbures). This development included major advances in deep-sea blow out preventer controls, re-entry sonar and various drilling tools. This project has led directly to the ability of Community companies to drill in up to 1,500 m water depths.

In <u>deep sea production</u>, studies were carried out by GERTH which led to the construction of the GRONDIN underwater test station. This was the first test made on a live field of a totally underwater production system designed for deep water.

- <u>1976 - 2nd Round Projects</u>

In <u>deep sea drilling</u> support was given to develop several novel aspects of the dynamically positioned drill ships "Pelerin" (GERTH) and "Ben Ocean Lancer" (BEN ODECO). The development of these two vessels added considerably to the ability of Community companies to drill for hydrocarbons in a deep water hostile environment.

The second round of projects marked the beginning of interest in novel <u>production systems</u>. As a result of work done in this round one Community contractor, VICKERS OFFSHORE, has been involved in the design of the world's first tethered buoyant platform for the HUTTON field in the North Sea. Also, another Community contractor, BRITISH PETROLEUM, has been chosen by the Norwegian state oil company STATOIL to develop deep water gas production technology.

Excellent results were obtained in the area of <u>natural gas technology</u> and Community support has succeeded in putting PREUSSAG AND SALZGITTER in the forefront of the LNG developments.

This round also marked the emergence of enhanced recovery technology, and a very successful pilot project was carried out by GERTH on the CHATEAURENARD reservoir. If the results of this project prove applicable on a wide scale, the recoverable reserves of the Community will be substantially increased ; estimates of the increase range from 10 to 50 million m^3 of oil.

- 1977 - 3rd Round Projects

In production technology a major project carried out by GERTH has led to the development and testing of components for a deep water production system. Every aspect of the system has been brought to a stage where a suitable field is now being sought on which to confirm the results of component tests. A valuable offshoot of this programme has been the further development of the J-CURVE pipelaying method, which is of major importance.

PREUSSAG continued to develop their experience in gas technology by carrying out a comprehensive examination of offshore liquefaction of natural gas. This project will have many applications in North Sea marginal field development.

Three enhanced recovery programmes were successfully completed by AGIP, BRITISH PETROLEUM and GERTH. These programmes not only give European oil companies hands on experience of this very important technology but should, in the medium term (5 years), lead to an increase in recoverable reserves. Estimated range from 20-100 million m^3 of oil.

A novel pipeline compression station was successfully developed by BORSIG ; this system has been installed on a test basis on a pipeline and is proving very successful.

- 1978 - 4th Round Projects

A novel drilling technique was developed by GERTH for drilling horizontal wells. This technique has proved very successful and will be used shortly to develop the Rospo Mare field in Italy.

Production technology was very much to the fore in this round ; an insert wellhead was successfully developed by SHELL INTERNATIONAL and is presently under test. BRITISH GAS have successfully developed a fluidised gas bed which has already attracted the interest of some Japanese gas companies.

The enhanced recovery sector grew for the third successive year and although some of the projects have yet to be completed, in general the results have been very encouraging.

In pipeline technology, a project to develop a cryogenic flexible by COFLEXIP is about ato attain its objective. This development will have enormous potential, especially in the areas of offshore loading of liquid natural gas and various natural gas liquids.

DIFFUSION OF KNOWLEDGE AND RESULTS

The diffusion of knowledge and results of projects in the Community Hydrocarbons Project Scheme is mainly the perogative of the contractor. Where results have been positive they have been presented as papers in technical conferences or have been diffused by means of public advertising.

The Commission has aided the diffusion process by two actions:

a) In April 1979 a Symposium was held in Luxemburg at which the result up to that time were presented and discussed by Community contractors and other interested oil and gas industry representatives.

b) The Commission regulary holds co-operation meetings, these meetings have a dual purpose, firstly they encourage co-operation between Community contractors working on similar problems and secondly they act as a forum for the presentation of results and the discussion of future measures to be taken.

INFORMATION FOR FUTURE PROPONENTS

INVITATION TO TENDER-SUBMISSION OF PROPOSALS
==

The Commission normally seeks to attract research and development proposals for subvention by the Community by issuing an annual call for tender. Since the provision of subvention may interest a great many proponents, the call for tenders is published in the Official Journal of the Communities.

It has been the practice to publish the call for tenders in September with a closing date for proposals of November 30th.

Research proposals should be submitted in the form required by the call for tender with particular attention being paid to the number of copies required.

Proposals should be sent to:

MR. G. BRONDEL
Director Hydrocarbons
Directorate General for Energy
200, rue de la Loi

B - 1049 BRUXELLES

Further information on the programme may also be obtained from this address.

TABLE 1

BREAKDOWN OF PROJECTS BY SECTOR

	Exploration	Drilling	Production	Enhanced Recovery	Transport	Storage
1st round (1975)						
Subvention (in EU)	432.000	9.380.000	13.790.360	1.883.673	11.945.926	1.482.000
N° of projects	(1)	(2)	(6)	(2)	(4)	(1)
2nd round (1976)						
Subvention (in EU)	3.287.933	1.483.200	20.913.651	7.358.101	1.353.500	--
N° of projects	(3)	(1)	(18)	(5)	(3)	(-)
3rd round (1977)						
Subvention (in EU / EUA)	605.344	2.160.528	29.376.876	3.399.101	5.663.303	290.246
N° of projects	(4)	(1)	(19)	(3)	(4)	(1)
4th round (1978)						
Subvention (in EUA)	1.055.450	4.281.599	19.517.390	4.881.513	5.091.159	1.864.769
N° of projects	(5)	(4)	(21)	(5)	(5)	(3)
5th round (1979)						
Subvention (in EUA)	354.059	--	10.996.650	2.082.395	9.323.705	158.875
N° of projects	(2)	(-)	(10)	(2)	(8)	(1)
6th round (1980)						
Subvention (in EUA)	1.732.299	--	16.434.162	6.331.584	1.749.120	1.849.999
N° of projects	(4)	(-)	(21)	(5)	(5)	(1)

1975

1st Round Projects

Title : System of Acquisition of Seismic Data — SN 348	Project N° : 1/75
Contractor :COMPAGNIE GENERALE DE GEOPHYSIQUE/SERCEL Address : Rue de Bel Air Carquefou — F 44470 Nantes Technical director (or person to contact for further information) : M. Hythier	Telephone N° : 40491181 Telex : 710695 CARQF.

This advanced seismic data acquisition system has been designed to meet the demand for more recording channels. Today, instruments can record 100 channels but multi-conductor cables and their plugs are a source of time consuming and costly maintenance. Recording 100 channels or more, as required by newly developing methods, calls for the use of Telemetry systems.

Telemetry provides the advantage of eliminating the transfer of weak analog signals over lenghts of conductors where they are polluted with distorsion and superimposed noise. In the SN 348, this advantage is pushed to maximum by plugging one geophone group only per station unit. The seismic signal is processed and digitized where it is generated. The damping is the same for every geophone group and the source impedance is the same for each channel.

The digital information is carried over a limited distance only, from a geophone station to the next, and regenerated by special repeater circuits in each unit until it reaches the Central Control Unit where it is recorded. With enough station units at 100 m intervals, there is no limitation to the length of the spread and transmission is not impaired by adverse environment, nor by leakage.

Early 1977, the first system is actually being tested on an experimental basis with 48 recorded traces. A second system is presently bench tested and the assembly of more field units continues.

The system has now been perfected and the project is completed.

Title : Deep—water Drilling	Project N° : 2/75
Contractor : GERTH Address : 4, Av. de Bois—Préau 92502 Rueil Malmaison — Paris Technical director (or person to contact for further information) : M. Leblond	Telephone N° : 749.02.14 Telex : 69066 F

In 1976 the project concerning Deep—water drilling examined several of the technical problems, the principal results are outlined below:

Classical mooring

Funicular anchoring is feasible in 1.000 m of water on condition that anchor scraping resistance can exceed 200 t.

Dynamic Positioning

A system utilising a sonar doppler can be used which adjusts for changes in position during deep—water drilling.

Risers

The computer programme simulation of the dynamic behaviour of the riser was compared to computer programmes of several oil companies.

Float modules

At the conclusion of the first step it seems that the buoyancy materials provided by Emerson Cummings and SNPE are suitable for 1000 m depths.

Riser emptying

A study to evaluate phenomena in the riser following a sudden dis-connection was carried out.

B.O.P.

The programme "Deep—water BOP" was followed with a study carried out to permit a better utilization of sonar and adjusting b.o.ps.

Measurements

Measurements on the foot of the riser necessitated the construction of a prototype which provided information on the fundamental parameters.

Re-entry sonar

Harmonisation of the multiplex transmission of television and re-entry sonar.

Underwater and retrievable plugs

Study completed.

Title : The Dynamically positioned Drillship "Petrel"	Project N° : 3/75
Contractor : OFFSHORE EUROPE S.p.t. Address : 113, Begijnenvest Anvers Technical director (or person to contact for further information) : M. Deckers	Telephone N° : 379950 Telex :

A. Objectives

To make operational for 600 m water depth the following devices:

(1) Dynamic positioning
(2) Riser and associated equipment
(3) Re-entry and reconnection
(4) Diving and underwater intervention
(5) B.O.P. handling
(6) B.O.P. control
(7) Special drilling equipment

B. Results

Preliminary remark:
The effects on the water depth and the corresponding water pressure on the equipment has not been established practically.

For water depths up to 300 m it is possible to judge the devices as follows:

(1) Dynamic position : good to very good
(2) Riser and associated equipment : good
(3) Re-entry and reconnection : good
(4) Diving : average to good
(5) BOP handling : good
(6) BOP control : not used (except for tests).
(7) Special drilling equipment
    - Ram compensator : to be finalised
    - other equipment : very good

Title : Deep Water Oil Production	Project N° : 4/75
Contractor : GERTH	Telephone N° : 749.02.14
Address : 4, Av. de Bois Préau 92502 Rueil Malmaison — Paris Technical director (or person to contact for further information) : M. Leblond	Telex : 69066 F

The production facilities available had essentially been designed for working oilfields lying below the continental shelves.

Their extrapolation for use in zones covered by a depth of water of from 200/300 to 1.000 metres was impossible, except for undersea wellheads, the industrial use of which was only in its infancy.

Analysis in 1974 of the status and medium—term evolution of off—shore techniques, together with an examination of the many conceivable systems for harnessing oilfields in deep waters, guided the project towards the study of the main components common to the most realistic systems, based on maintaining the greatest possible proportion of installations on the surface.

Firstly, this work led to the elaboration of a system of engineering files for a certain number of components, supported by mathematical simulation programs and testing of reduced—scale models in the testing tank and secondly, the conduct of several tests confirming the validity of the solutions adopted.

Among the components examined were:
a) Articulated columns.
b) Submerged storage at great depth.
c) Anchored floating supports and production risers.
d) Undersea pipelines.
e) Oil well servicing.
f) Appurtenant techniques.
g) 70 m depth production pilot installation — Grondin.

This contract was completed on 31.12.1977 and a complementary work programme is being undertaken in contract 03.35/77

Title : Diverless Flowline and Pipeline Connections	Project N° : 6/75
Contractor : SEAL Ltd. Address : Technical director (or person to contact for further information) :	Telephone N° : Telex :

Introduction

The production of hydrocarbons from great water depths (more than 1,000 ft) by subsea production systems needs multiple or single connections between the sea bed located structures corresponding to these systems. In the water depths under consideration, these connections have to be made without diver intervention.

Objectives and scope

As it is an interface problem, the connection between sea bed located structures can only be considered in particular cases. Nevertheless, the study of the following particular problems allows for the definition of enough specific solutions to answer most of the questions which are subject to occur:

- single line at arrival or departure from a structure, when bottom shifting and pulling-in are possible;

- single line associated to a cluster:
  - internally: connection between the trees and the gathering manifold
  - externally: connection between the gathering manifold and the production support;

- bundle of lines between manifolding structure and gathering point.

The objectives of the programme concerned with the contract 6/75 are as follows:

- to define the techniques and equipment which enable the above described connections to be made;

- to demonstrate the feasibility of the so defined methods.

Results

As the programme is underway, the results which can be described are intermediate and concern the discussions and choices made in the course of the studies. The following results are of particular interest for each application defined above:

- single line connection:
  - beginning of laying: remote connection and laying of a flexible line.
  - end of laying: laying and remote connection of a rigid line by a reel barge;

- connections associated to a cluster:
  - internally: by a remote machine positioned on rails (under study)
  - externally: laying of a rigid line by towing it out near the sea bottom (stabilization by chains and floats) and remote connection after out of an overlenght and tooling on the sea bottom by a remote controlled machine;

- bundle of lines: towing out, on the surface, of a bundle fabricated onshore and equipped with floats, and laying on site by drawdown and adjustment of the flotation. The connections are made by remote controlled connectors positioned on the bundle ends.

Conclusion

The study of the various problems defined above has shown that, in all cases, the key problem is the interface between the laying of the line and the second connection. This has determined the choice of the global solutions as described below:

- use of a flexible line allowing, by deformation, for adjustment, in angle, in length and in position, of the end of the line;

- use of a rigid line, in conjunction with a laying method allowing for an axial positioning, and adjustment in length by remote cutting on the sea bottom;

- use of a rigid line, in conjunction with a laying method allowing, by elastic deformation, for an axial positioning and an adjustment in length.

Title : Automated Subsea Wellhead	Project N° : 10/75
Contractor : TECNOMARE	Telephone N° : 708622
Address : San Marco 2091 30124 Venezia – Italy Technical director (or person to contact for further information) : Ing. Banzoni	Telex : 41484

Scope of the project was the design and construction of a prototype of an automated underwater wellhead.

The design and construction of a complete remote control system which utilizes three different transmission links (acoustic transmission through the sea water, electrical cable and electrical transmission via flowlines) has been carried out.

The remote control system includes, in addition to the electronic subsystems, the subsea electro hydraulic plant.

The system is designed to operate up 600 mt and at a distance from the terminal platform of 8 km.

The work done has led to the following main results:

a) Construction and sea trials of a laboratory of the electronic subsystems based on the transmission of acoustic signals through the sea water. The system has proved its capability for a distance of more than 8 km.

b) Construction of the operative prototype of the electronic sub-system. Design and construction of the subsea electro-hydraulic plant and of all the auxiliary equipment.

c) Sea trials of the whole system in 70 mt with transmission both via cable and via acoustic propagation. The positive results of these tests have demonstrated the validity and reliability of the control system.

Title : Research on Oil Recovery from Heavy Oil Recovery from Heavy Oil Deposits Under the Adriatic Sea	Project N° : 11/75
Contractor : AGIP Address : 20097 S. Donato Milanese 20100 Milano — Italy Technical director (or person to contact for further information) :M. Chiarichi	Telephone N° : 53531 Telex : 31246

After the 1973 oil crisis, AGIP and Deutsche Shell embarked on an EEC—subsidised joint venture, in order to further explore the possibilities of developing the heavy oil of the fractured limestones of Upper Cretaceous-base Teriary ages of the Adriatic Sea area offshore Pescara. The work for the first phase consisted of the following parts:

— Planning the drilling, coring and testing of a well in the Emilio structure and carrying out such operations.

— Analysis of cores from the well (Emilio—4),

— Laboratory study to initiate the development of a suitable recovery method.

Well Emilio—4 was spudded in on June 3, 1976 and reached the final depth of 3400 m on September 20, 1976. It was continuously cored from 2890 to 3240 m and a final core was taken from 3444 to 3450 m. The cored section was found practically tight with a few fractures only, showing stains of very heavy oil. The matrix was found to be water—impregnated. One production test and seven drill—stem tests were performed with disappointing results.

The cores and fluids recovered were studied in the Koninklijke/Shell Exploratie en Produktie Laboratorium; the following results were obtained:

1. Although Well Emilio—4 was found non—productive, data from this well have been valuable in contributing to the knowledge on Adriatic Sea heavy—oil carbonate reservoirs in general and to the evaluation of Emilio field prospects in particular.

2. The geological analyses have revealed seemingly exceptional oil accumulations brought about by intrusion of immature heavy oil from a source rock. As a result, the oil has saturated sparsely distributed porous carbonate hones. Furthermore, it has fractured the more abundant tight rock, and from these fractures the oil has to some extent penetrated the tight rock matrix.

3. Although the oil-intrusion model under 2 does not permit the oil
   distribution to be predicted very accurately with existing petroleum-
   geological methods, it is most likely that the actual oil-in-place is
   even less than 25 percent of the original estimate.

4. The typical oil occurrence as described under 2 and 3, the much lower
   quantities of oil than originally foreseen and the special well
   completions required to bring the oil to the surface, leave no scope for
   economic oil recovery. It is therefore recommended that further evaluation work on developing practical oil-recovery method(s) for the
   Adriatic Sea deposits be discontinued.

Title :Acquisition of a Production Technique for the Exploitation of Deep Sea Deposits of Hydrocarbons	Project N° : 12/75
Contractor : L.E.A.	Telephone N° : 26140
Address : c/o W.S. Atkins Woodcote Grave, Ashley Rd, Epsom — Surry KT185BW Technical director (or person to contact for further information) :M. Miclethwaite	Telex : 23497

The work on all parts of the L.E.A. study was completed by December 1978. The progress with each part was as follows:

1. The study into the sensitivity of steel jackets, lack of knowledge of input properties has aroused much interest as a means of guiding research both in the general sense and also with respect to specific jacket designs.

2. The computer programme ASALAUNCH has been developed and is being used to check the launching of a number of steel jackets.

3. The work on water-structure interaction is being published at several conferences and the techniques developed have been used by other EEC grant holders.

4. The design for a jacket for 250 m water depth is complete and lessons learnt from it and particularly the method of launch have been translated to other designs.

5. The integrated deck scheme has been offered to several oil companies and is currently being assessed by them.

Title : Remote Controlled Subsea Handling Vehicle	Project N° : 13/75
Contractor : WINN TECHNOLOGY	Telephone N° : Bandon 49601
Address :Kilbrittain co. Cork	Telex : 8443
Technical director (or person to contact for further information) :	

This contract was completed in August 1979 as per the contract and culminated in the production of an operational machine capable of:

- manoeuvering in any direction on the seabed under control from a console on the surface, navigation being accomplished by resolving direction and distance information received from the on-board gyro compass and wheel rotation. The resulting components are indicated on a chart at the control console.

- Manipulating objects up to 250 lbs. in weight at the extremity of the arm which is fitted with a manipulating claw, developed under contract 07.88/76

   The claw, which is not removable, can in turn, pick up additional devices developed under contract 07.88/76.

- Performing operations such as valve turning and the unscrewing of nuts, bolts and clevis pins.

   The launch and recovery gear unit is skid mounted complete with its power generator and umbilical cable reel. The console is mounted separately.

14

Title : Underwater Oil Storage Tank	Project N° : 15/75
Contractor : **TECNOMARE** Address : San Marco 2091 30124 Venezia — Italy Technical director (or person to contact for further information) : Ing.Rodighiero	Telephone N° : 708622 Telex : 41484

The main targets of underwater oil storage tank research project are described in the following:

— to obtain and/or transfer the basic methods and techniques to the design of large offshore structures made of steel or of reinforced concrete

— to develop the design of a 300 m water depth underwater oil storage tank.

The design of the underwater oil storage tank derives from the integrated solution of different problems regarding:

Storage process system, structure configuration, design methods, installation procedures, construction methods, operative life and maintenance procedures. Different design aspects have been analysed in detail so that computer procedures, theoretical studies, model tests and field tests have been carried out.

The results of the experimental test performed confirm the feasibility of the concept with special reference to the fabrication, insolation and operating procedures.

The system of underwater storage oil tank has a configuration flexible enough to meet a wide range of storage capacities, installation water depths, metoceanographic conditions and types of connection production and mooring systems.

It represents a valid solution for exploitation of marginal fields over 100 m and for large fields in deep water where floating storages become uneconomical.

Title : Research on the Uses of Oil Reservoirs in Fractured Rocks for the Storage of Liquid and/or Gaseous Hydrocarbons	Project N⁰ : 16/75
Contractor : AGIP	Telephone N⁰ : 53531
Address : 20097 S. Donato Milanese 20100 Milano— Italy Technical director (or person to contact for further information) : M.Chiarichi	Telex : 31246

Since 1957 Gela field, Italy, has produced very heavy and viscous oil from a reservoir at a depth of 3,200 to 3,500 metres. The reservoir oil is undersaturated and shows both an areal and vertical variation of its characteristics. The reservoir geometry is complicated because of the presence of a system of faults and the variation of the degree of fissuring throughout the field. Production mechanism is by water drive; the reservoir is rate—sensitive and the oil production rate is kept at a level low enough to prevent water production problems. The purpose of the research was to find out whether oil recovery can be enhanced by replacing the natural water drive mechanism with the injection of either natural gas or carbon dioxide.

Laboratory tests have been performed on cores to assess how gas drainage improves oil recovery in comparison with water imbibition, and how injected gas affects the thermodynamic characteristics of reservoir oil. A three—dimensional, black oil numerical model has been employed to evaluate the improvement in oil recovery and oil production rate that can be achieved under different gas injection schemes. The model had been previously validated by matching field past history; the future field performance by natural water drive has also been evaluated, to provide a basis for comparison.

Title : LNG Storage in Undergrounds Caverns	Project N° : 17/75
Contractor : GEOSTOCK	Telephone N° : 7785353
Address : Tour Aurore Cedex 5 — 92080 Paris la Défense Technical director (or person to contact for further information) : M. Faucon	Telex : 610898

A methodology for building underground storage facilities for low temperature liquefied fluids has been outlined from the research described. The methods so developed provide solutions to the problems encountered by previous unsuccessful attempts, and GEOSTOCK will soon be in a position to design full scale cryogenic LNG and LPG underground storage facilities once the pilot cavity has been built to finalize them. Patents for the processes involved have been applied for.

The technical and economic considerations described in the foregoing lead to the following conclusions on which the remaining development will be based:

a) Underground dryogenic storage is still a thoroughly attractive proposition despite temporary political and/or economic opposition,

b) a smaller pilot cavity can be substituted for the large section originally considered to keep down construction time and cost, and can also be used for investigating LPG as well as LNG storage, and

c) in the present economic context, the cost of this trial is only justifiable if there is a good probability of the result being applied to commercial storage facilities.

Since a commercial requirement for this type of storage has not materialized it was decided to terminate this project.

Title : Deap Sea Pipeline	Project N⁰ : 18/75
Contractor : GERTH	Telephone N⁰ : 7490214
Address : 4, Av. de Bois Préau 92502 Rueil Malmaison — Paris Technical director (or person to contact for further information) : M. Leblond	Telex : 69066 F

1973 marked a turning point in the history of transportation of oil and gas by undersea pipelines, since it was in 1973 that a series of developments started in the North Sea with the construction of large diameter conduits (32 inches – 34 inches) several hundred kilometres long and in depths of about 150 m.

With difficult sea conditions combined with technical difficulties, the only solution for the problems encountered was to replace more or less empirical methods with rigorous methods resulting from exhaustive analysis of all the problems arising, coupled with the use of higher performance equipment and well adapted working procedures.

To meet these requirements, the DEEP STEEL PIPELINE project encompassed all the problems involved in the laying, repair and use of undersea pipelines for diameters of up to 40 inches, in difficult conditions of sea and for depths of up to 500 m.

The two principal objectives of the project were:

– to improve existing means and techniques and, if necessary, extend their range of use,

– to study and develop new methods, allowing for the operational or economic limits of these techniques.

Conclusion

The development and application of the dynamic computation programme turned out to be a very delicate matter and this programme cannot be considered as operational.

The studies performed on anchoring systems and the equipment of the laying barges have enabled the depth limits to be defined for the types of tubes to be laid to the following conclusions:

– for heavy barges, laying of large diameter tubes (30'') is possible down to a depth of 400 m. This depth can be increased to 600 m by modifying the anchoring device,

— for light barges capable of being anchored dynamically, the laying capacities are:

a) a 30'' pipeline in x 65 steel down to a depth of 300 m (density — 1.08)
b) a 12'' pipeline in x 65 steel down to a depth of about 850 m.

No applications were found for the anti—heave system for handling heavy packages owing to its complexity and size.

To conclude, these studies have shown that it was possible to use conventional laying means in depths of water of 400 to 600 metres.

The logical outcome of this project could well have been a laying test in a depth of 400 to 600 metres using these means. Owing to the work conducted in this area in the Mediterranean (crossing of the Straits of Messina), it was decided not to carry out this test, which would have repeated other work, and to bring most of our efforts to bear on a new laying technique (the R.A.T. method indicating towing, end connection and tensioning) capable of optimum performances both with respect to rate of advance and laying depth.

A complementary programme of work is being undertaken in contract No. 09.07/77.

Title : General Study and Field Testing of Deep Water Submarine Pipelines	Project N° : 19/75
Contractor : TECNOMARE	Telephone N° : 708622
Address : San Marco 2091 30124 Venezia – Italy Technical director (or person to contact for further information) : M. Rodighiero	Telex : 41484

The scope of the project was the study and the field testing of the laying of pipelines in deep waters and was constituted by the definition and the development of the technology to be followed to reach pipe laying capability in water depths up to 1000 m in connection with pipe sizes ranging from 6'' to 48''. The main technical results were the following:

a) Computer procedures. The set up calculation methods allow to analyse all the principal conditions the laying equipment may meet.

b) Model tests: Model tests have been performed on the different components (barge, stinger, mooring line, pipe etc.) and on the complete laying system for analysing the behaviour of the stinger pipe system and the moored lay barge.

c) Field tests: The experimental data collected during the sea trials carried out by SNAM in the Sicily Channel have been analysed and compared with the theoretical results derived from the computer procedures.

Both the model and field tests results confirmed the validity of the computer programmes.

d) Study of new techniques and equipment: Many techniques and equipment necessary to solve specific problems encountered during pipe laying operation in deep waters have been pointed out and studied.

Title : Laying Tests in the Straits of Messina	Project N° : 20/75
Contractor : SNAM S.p.A.	Telephone N° : 53531
Address : 20097 S. Donato Milanese 20100 Milano – Italy Technical director (or person to contact for further information) : M. Cicarelli	Telex : 31246

The tests concerning the above mentioned SNAM's Project had as their scope to verify the feasibility of laying down a pipeline in deep waters up to 360 m. with heavy environmental conditions such as:

– difficult geomorphological structures
– strong and quickly variable sea currents.

Furthermore the tests had the aim to experiment with new measuring and control systems to verify the accuracy of the mathematical models utilized in the computer software, to specify the technical limits of the laying procedure and the possibility to coordinate laying operations by submarine assistance.

For said test a pipe having the following characteristics was utilized:

– diameter 10'' (27.3 cm)
– thickness 15.88 mm
– material according to API Std 5LX-X52
– polythene external coating.

The main conclusion is that the employed techniques allowed the laying of the above-mentioned pipe up to the maximum depth of 360 mt.

The conditions of the sealine are satisfactory, for both its integrity and its static support on the sea bottom and this was largely proved by the pressure test and by the many inspections made by submarine.

Finally the utilized equipments showed to be suitable and convenient for the pre-established scopes.

Title : Laying Tests in the Sicily Channel	Project Nº : 21/75
Contractor : SNAM S.p.A. Address : 20097 S. Donato Milanese 20100 Milano — Italy Technical director (or person to contact for further information) :M. Cicarelli	Telephone Nº : 53531 Telex : 31246

The tests concerning the above-mentioned SNAM's Project had the scope to verify the feasibility of laying down a pipeline in deep waters (up to 550 mt), focusing on the relevant laying procedure and acquiring at the same time a wide range of data.

In order to have the largest possible amount of information, it was decided to lay two different sections of pipeline (12'' and 16'') having a length of 3.2 km and 3.5 km respectively.

Both the sections have been laid down at the maximum depth and in such a way to cross flat and uneven areas.

The other main characteristics of the pipes used are:

- thickness (16'') 17.48 mm
- thickness (12'') 15.88 mm
- materials according to API Std 5LX-X52 and to special specifications
- external coating (for testing only) selected in the following types: polythene, neoprene (2 kinds), epossidic resin.

All tests have already been completed and, it is possible, since now, to state the most important conclusion which is that the employed techniques allow the laying of the above mentioned pipes up to the maximum depth of 550 mt.

The conditions of the sealine are satisfactory, either for its integrity as well as for its static supporting on the sea bottom and this as largely proved by the pressure test and by the many inspections made by submarine. On the other hand it is appropriate to point out that the utilized laybarge showed to be, as foreseen, particularly sensitive to the metheorological conditions, taking also account of the necessary routing orientation practically perpendicular to the direction of the most dangerous storms.

This, if from one side caused some problems, allowed to gain those experience and data to improve all the employed equipments.

1976

2nd Round Projects

Title : Study of Particular Problems Involved in Seismic Prospection	Project N° : 01.02/76
Contractor : GERTH/CGG	Telephone N° : 7490214
Address : 4, Av. de Bois Préau 92502 Rueil Malmaison — Paris Technical director (or person to contact for further information) : M. Leblond	Telex : 69066 F

This project was jointly carried out by GERTH and CGG.

- Deployment of the Flexichoc source.
  The apparatus of the multiple flexichoc 506 permitting the deployment of sources is operational.

- Large profile utilizing Flexichoc source.
  A series of experiments were carried out, results were analysed during 1979.

- Large profile utilizing the vaporchic source.
  Testing of the Coflexip flexibles have been completed. It appears that this solution requires additional study.

- Geometric control of sources and 'fish'.
  Tests have been carried out, they have shown the system of geometric control of sources and fish is feasible.

- Integral Navigation.
  Tests carried out at the beginning of 1979 were judged unsatisfactory and a re-estimation of the performance one can expect to attain with the system studied showed that it will be difficult to obtain system performance in navigation.

Title : Research on the Methodology and Geophysical Techniques Applicable to Particularly Complex Geological Situations	Project N° : 01.03/76
Contractor : AGIP	Telephone N° : 53531
Address : 20097 San Donato Milanese 20100 Milano — Italy Technical director (or person to contact for further information) : M. Chiarichi	Telex : 31246

The programme follows up a previous three years of research and development carried out under this project. A large volume of seismic data has been collected in order to improve the reliability of the seismic information in areas of complex geology in connection with severe surface conditions.

During 1980 the work was essentially devoted to a series of experimental tests on Apulia region. Apulia, a large area located at the extreme south—east edge of Italy, is a wide and thick carbonatic platform. A few seismic lines recorded in the past didn't show any results so far a sequence of tests was provided to give an ultimate answer on the best field and processing techniques to be used. The hydrocarbon targets are at great depth (more than 6000 metres) then major efforts have been focused to get deep seismic information. The tests were carried out by a seismic crew equipped with a 96 standard channels telemetric unit. Afterwards, in order to improve the performance, the number of channels were extended to 144.

A second crew equipped with heavy vibrators carried out a test line for comparison.

Main objectives of the tests were as follows:

— Analysis of the propagation noises and capability of their attenuation by field techniques and processing procedures.

— Analysis of the source signatures.

— Spectral analysis of the signals.

— Feasibility and performance of the shot—in—air method (POULTER).

— Comparison of the results among: poulter, deep hole shooting and vibrators.

On the northern part of the area a wide extended spread was surveyed with a multiple coverage up to 96 fold stack.

Present work is mainly devoted to the processing of the recorded data. The attempt is to get good results by the best compromise among field techniques, processing and costs.

Title : System of Medium and Long Range Distance Measurements	Project N° : 01.04/76

Contractor : CGG/SERCEL
 	
Address : SERCEL
     Av. de Bel Air — CARQUEFOU
           NANTES
Technical director (or person to contact for further information) : M. Hythier
	Telephone N° : 40491181
	
	Telex : 710695 CARQF

Merops is a precise and non-ambiguous system of radio location which operates in the V.H.F. band and which has been developed to aid the positioning of ships, barges or platforms. It constitutes an extension of the SYLEDIS system which has been widely used in offshore operations since 1975.

Since, SYLEDIS is limited in range to between 100 and 150 km. The techniques developed by MEROPS assume a positioning precision of 10 – 20 m at a range between 250 – 400 km, night or day and in all types of weather. This accuracy is obtained despite the considerable adverse conditions which the wave propagation encounters at the horizon: large attenuation, fluctuations in level and propagation time, effects of abnormal refractions.

The MEROPS system embodies:

— A power amplifier (20 – 320 watts).

— A filter section which excludes all non essential components of the spectrum (satisfying the most stringent national regulations: level of 5 microwatts out of band.)

— A series of "anti-refraction" antennas equiped with a means of electronic commutation.

— A complementary means of statistical filtration of position data (U.C.M. and filter).

A permanent chain with large coverage is presently being developed in the Gulf of Mexico (from Louisiana to Mexico).

Title : Development of Deepwater Drilling Technology	Project N° : 02.06/76
Contractor : BEN ODECO LIMITED	Telephone N° :031-2252622
Address : 29, Bernard St. Edinburgh EH66R4	Telex : 72611
Technical director (or person to contact for further information) : J. Tolson	

Ben Odeco Ltd. is a British Company which operates the jack-up rig OCEAN TIDE and the conventional drillship BEN OCEAN TYPHOON.

In 1974 Ben Odeco Ltd. (B.O.L.) signed a contract with Scotts Shipbuilding Co. Ltd., shipbuilders of Greenock in Scotland, to build a dynamically positioned drillship based on an IHC design for which Scotts held licence rights, but to be developed to drill in 3000 feet of water. This drillship named BEN OCEAN LANCER was delivered to BOL in March 1977.

The principal dimensions of the drillship are:

- Length BP   135.8 metres
- Moulded depth   23.45   "
- Moulded depth   12.45   "
- Design draft   8.00   "
- Max. load in salt water at design draft   8.228 tonnes
- Displacement in salt water at design draft   17.728   "
- Speed about   13 knots

To achieve the drilling systems required to operate BEN OCEAN LANCER in water depths as deep as 3000 feet (915 metres) B.O.L. has undertaken research and development which has included the following specific studies:

a) the study of potential acoustic and associated problems
b) the study of marine riser in deep water and the specification of riser flotation and related systems
c) the design of blow-out preventer (BOP) multiplex controls for operation in 3000 feet of water
d) the interfacing of the reentry systems with dynamic positioning systems.

On the basis of these studies, systems have been designed, manufactured and installed in the rig. The effectiveness of the design systems will be evaluated over an extended period of operations during which it is anticipated that the vessel will drill in various water depths including wells at 3000 feet.

The vessel uses an acoustic beacon at seabed level as one of its positioning references. A detailed study has been conducted into the various ways in which the acoustic signal might be swamped by extraneous acoustic electrical or mechanical "noise", and a series of "noise" measurements (amounting to a noise signature of the vessel) has been recorded and is being analysed, to provide a comparative base to assist with the diagnosis of any future problems in this sphere.

The studies of the performance of marine risers in deep water has been used to design the flotation cladding for BEN OCEAN LANCER's riser. The final design of flotation material required the redesign of the drillfloor of the rig. A gimballed drillfloor design was adopted which allows the rotary table and diverter to maintain a steady position with regard to the riser irrespective of the motion of the vessel when the riser is being run.

The anticipated command distance for BOP controls when drilling in deep water and the absolute requirement for immediate BOP reaction has led to the development of an electro-hydraulic multiplex BOP control system. The system operates by means of a coded electronic impulse transmitted through cables to a control unit on the BOP stack which then works the rams hydraulically. If for any reason the rig has left the wellhead then acoustic back up controls can be used to operate certain functions on the BOP.

The wellhead reentry system is interfaced with the dynamic positioning system so that the vessel will automatically position itself over the wellhead once the re-entry system has located it.

Title : Development of an Automatic Mooring System	Project N° : 03.12/76
Contractor : WHARTON ENGINEERS	Telephone N° : 01.953.2205
Address : Watford Rd. – Elstree Boreham Wood – Herts WD 63 BT Technical director (or person to contact for further information) :	Telex : 922919

Phases 1 and 2 have been completed and the following have been established:

1. The requirement of a load sensing equipment with sufficient sensitivity to give the necessary response.

2. The magnitude of the forces acting on a vessel and identification of the rate of change of these forces.

3. Examination of the spring characteristics of various mooring systems at a variety of water depths to assess how much increment was required to change the force on the mooring lines and how quickly it should be carried out.

Phase 3 completed the following tasks:

1. Tank tests have been carried out in conjunction with B.P. Tanker Co.

2. The traction winch unit has been further developed.

The report on phase 3 concluded that little further work could be done until a system was sold.

Development work has been suspended and a suitable client is being sought.

Title : Exploitation of Oil and Gas Fields Using Floating Platforms	Project N° : 03.13/76
Contractor : TECNOMARE S.p.A. Address : San Marco 2091 30124 Venezia — Italy Technical director (or person to contact for further information) :M. Rodighiero	Telephone N° : 708622 Telex : 41484

The objective of the project was to investigate and solve the problems involved by the design, construction and installation of large offshore floating platforms for industrial plants in very deep waters. As a practical application of the research, a floating platform suitable for supporting a typical industrial plant was to be designed. In the project execution plan the work was divided into three main areas:

- Aquisition of know-how and tools for the design of floating islands. This included a number of theoretical studies on subjects like a study of the dynamic behaviour of offshore structures subject to vortex shedding and buffeting; study of the long term drift forces action on moored bodies, study of a general simulation method of weather conditions; etc. Computer programmes and calculation procedures were also developed for the analysis of the dynamics of floating framed structures; for the calculation of drift forces; for determining the downtime of marine systems, etc.

- Design of a typical floating platform for very deep waters and high payload. The design was specially aimed to an offshore oil and gas production application but the results are such to be readily applicable also for other types of industrial plants. After an analysis of different possible solutions the so called "tension leg" concept was chosen and several new solutions were found for design, fabrication and installation problems. The main specifications are the following:

  - water depth   600 m
  - weather conditions   typical of the northern North Sea
  - payload   26.000 t_2
  - deck area   10.000 m^2
  - gravity anchor bases
  - anchor lines made of steel pipes
  - flexible structural connection of anchor pipes to bases and platform

- Basic test of a platform model. Extensive tests were carried out on a 1:100 scale model of the platform in the Trondheim (Norway) ship model basin. The results were very satisfactory and showed a good correlation with the calculation results.

The project is now completed.

Title : Intermediate and Deep Sea Production	Project N° : 03.20/76
Contractor : VICKERS OFFSHORE	Telephone N° : 0229 27171
Address : Craven House Barrow-in-Furness – Cumbria LA14 1AF Technical director (or person to contact for further information) : T. Steer	Telex : 65197

1. Feasibility study

- Forecasts of offshore production for the northern North Sea and in Europe against a background of world energy supply and demand

- The potential demand and most likely applications for innovative engineering solutions such as the tethered buoyant platform

- The comparative economics of alternative development programmes for a range of fields and applications

- The identification of the most cost-effective innovative producing systems

- The engineering developments which will be necessary to bring about the acceptance of the TBP.

2. Initial design study

- The development of a suite of analytical techniques for predicting the dynamic behaviour of the tbp/moorings/riser system

- The development of techniques whereby a complete structural analysis could be undertaken of the tbp, taking account of all dynamic loading conditions

- Design studies on specific technical aspects, in particular methods of attaching and adjusting moorings, riser design and installation techniques

- Addition studies on technical areas of special interest.

3. Termination

The project terminated at the end of 1978.

Title : Research and Development Programme for the Production of Oil and Gas from Deep Water	Project N° : 03.21/76
Contractor : TAYLOR WOODROW CONSTRUCTION LTD. Address : 345 Ruislip Rd. Southall — Middlesex UB1 22X Technical director (or person to contact for further information) : J. Smith	Telephone N° : 01.578.2366 Telex : 24428

The main objective of this comprehensive design and research programme was the development of complete drilling and production systems for water depths beyond Continental Shelf limits. The systems were based upon the use of concrete articulated buoyant columns, termed ARCOLPROD, and the programme commenced in 1977.

The second stage of the programme, which was effectively completed in 1980, encompassed a detailed development of a reference design structure for a hypothetical North Sea field in 250 metre water depth. The structure was to be equipped for 120,000 BPD production, full drilling and nominal buffer storage. The design covered the complete structural and dynamic analysis, foundation considerations, construction method statements, plant and process design, safety analysis, pricing and field economic analysis. The complete design package and drawings received quality assurance from Lloyd's Register of Shipping, and the marine aspects were approved by London Offshore Consultants. Final reports and drawings were provided to sponsors and the five oil companies collaborating on the programme, and the follow-up presentations conducted in May 1980. All collaborators expressed satisfaction with the outcome of the development and recognized ARCOLPROD as a competitive system for deeper waters.

Major supporting research was also completed during the year. This included a provisional field study on the performance of tension piles to anchor the base of ARCOLPROD, and extended fatigue testing on the polyester tendons proposed for the articulating joint, both studies employing specially developed equipment.

The programme concluded with a sensitivity study of the system's performance in water depths up to 400 metres, and production capacities up to 180,000 BPD, confirming the technical capabilities and economic reliability.

Title : Development Project of a Novel Offshore Production System	Project N° : 03.27/76
Contractor :THE BRITISH PETROLEUM COMPANY LIMITED	Telephone N° : 01.920.8000
Address : Britannic House Moor Lane — London EC24 9BU	Telex : 888811
Technical director (or person to contact for further information) : I.M.Barrett	

The project involved the development of vertically tethered buoyant platforms and was executed in four phases. In addition to participation by the Commission of the European Community, participation agreements were signed for Phases 1, 2 and 3 with Vickers Limited (later British Shipbuilders) and with the Offshore Energy Technology Board on behalf of the Department of Energy.

Phase 1 consisted of a feasibility study conducted in late 1975 and early 1976 into the possible application of a tethered buoyant platform production system in the development of the Magnus Field.

Phase 2 consisted of an optimisation study on platform configuration and the tether and anchorage system, a feasibility study on the possibility of permanently installed risers in North Sea storm conditions and work to define the system components for a reliable cost estimate. This work was completed in late 1977.

Phase 3 was initially intended for the preparation of outline design for a specific application. Phase 4 would then develop a tender specification for a prototype installation. Since BP prospects for a specific application did not exist at the end of Phase 2 it was decided to separate Phase 3 into two parts.

Phase 3A involved further development engineering of aspects which were not field specific. Work was completed in 1979 and indicated that an increased scope of prototype testing was required. It was decided that since Phase 3B had been intended for the development of a specific field application which was yet to be established, the work would be covered by a re-designated Phase 4.

Phase 4 work commenced in late 1980 with the development of a testing programme. It was then decided by BP Management to recommend that further work be suspended pending a specific BP application. Work ceased in December 1980.

Considerable work had been executed during the earlier phases which confirmed a level of confidence that a vertically tethered buoyant platform was an economic and competitive option for deeper water application. Upon discovery by BP of a specific application, the outstanding Phase 4 objectives can be carried out concurrently with the overall design work.

Title : Mobile Offshore Natural Gas Liquefaction Plant	Project N° : 03.28/76
Contractor : PREUSSAG	Telephone N° : 0511 19321
Address : Postfach 4829 Arndtstrasse 1 – 3000 Hannover Technical director (or person to contact for further information) : D. Meyer-Detring	Telex : 922851

By end of December 1978 the work programme was completed with the exception of the approval in principle (sub-contract to the classification Society DET NORSKE VERITAS), the extensive hydrodynamic test programme with a scale 1:75 model of the floating plant (sub-contract to Versuchsanstalt für Wasservau und Schiffsbau, Berlin) and the design modifications resulting from the results of these two sub-contracts.

Meanwhile, the work programme was finished by 28th February 1979. The aim of the development has been reached. The floating LNG plant of the consortium 76 represents a technically feasible and economically attractive solution for the exploitation of marginal gasfields offshore.

Title : Production of Liquefied Natural Gas and Methanol	Project N° : 03.29/76
Contractor : SALZGITTER	Telephone N° : 05341-21-1
Address : Postfach 411129 3320 Salzgitter 41 Technical director (or person to contact for further information) :M. Holekamp	Telex :954481

In April 1975, a project group comprising LGA Gastechnik GmbH, Howaldts-werke-Deutsche Werft AG and Salzgitter Industriebau GmbH started preparing a documentation dealing with offshore processing plant for economic utilization of gaseous hydrocarbons from marginal fields in the North Sea. The project group was managed by Salzgitter AG, and the project was subsidised by the Commission of European Communities. Project work was completed in May 1978.

To exploit these gas fields, a natural gas liquefaction plant (18,500 m³ LNG per day) and alternatively a methanol processing plant (3,000 t/d) were designed. A fixed jack-up type platform was selected to serve as a supporting structure for both types of plant. This supporting structure which will also accomodate integrated buffer storage tanks, quarters etc., can be used together with its substructure for water depths up to 180 m.

Compared with floating offshore systems, this concept has various advantages. As a fixed platform is not subjected to motions, the process flow will be trouble-free. Furthermore it is not necessary to use flexible high-pressure risers that have not yet been tested and proved, where the gas is to be transferred from the bottom of the sea; and motion-compensated LNG lines to the sea bed or submerged floating lines that are required in the event of indirect loading need not be used. Finally the jack-up technique permits changing locations without major difficulties.

Both types of processing plant can be equipped with workover facilities so that well servicing work will not require any relocation of the platform.

To bridge bad weather periods during which loading onto vessels is not feasible, an integrated buffer store has been provided for intermediate storage of the product.

Product transfer is carried out externally via a buoy that is connected to the processing plant by an subsea pipeline. Considering the expenditure involved in an external transfer system, a follow-up project deals with developing a direct transfer system which would make it possible to do without such an external system. A product carrier suited to this new concept has already been designed.

The development work for an offshore methanol plant was completed by preparing detailed design documents for the entire system. A "Quality Assurance" has been granted by Germanischer Lloyd.

Project work on the large offshore natural gas processing plants has reached a similar stage. Further activities are concentrated on developing TLP versions for use in deeper waters, and on smaller plants for utilization of associated gas.

The processing systems developed by the Salzgitter project group were presented in various conferences and meetings with potential users. The discussions showed that these potential users considered this concept to be highly realistic in view of its possible implementation, especially because normally such a new technology would involve considerable risks that are reduced to a minimum in the selected concept. For the liquefaction process e.g. the Air Products system was chosen which is well-proven all over the world. Thus, when using a fixed platform, it is only necessary to adapt conventional land installations to offshore conditions. In a similar way, a number of peripheral components must only be modified. Accordingly, new developments can be restricted to an absolute minimum.

Based on this project and various supplementary development activities, the Salzgitter project group is in a position to submit technically and economically sound offers to potential users of this technology.

Title : Offshore Production System "Exboy"	Project N° : 03.31/76
Contractor :FREEMAN FOX AND PARTNERS	Telephone N° : 2228050
Address : 25, Victoria St. (Smith Block) London SW1 HOEX Technical director (or person to contact for further information) : Dr. Browne	Telex : 916018

    The project follows up an earlier two year research and development programme carried out in 1973-1975 on tethered bouyant platforms for use in exploiting undersea resources, particularly in deeper waters. Early work resulted in a design concept covering a wide range of payloads up to 22,000 tons and a wide range of water depths. As a basic marine platform the design can be used or adapted for any purpose connected with offshore oil or gas production.

    The project is essentially devoted to the platform structure and its hydrodynamic behaviour, together with its associated anchorage systems, and the present phase takes the development of such a platform to a point where the design is suitable for commercial application.

    In 1977 and 1978 work was carried out on optimising alternative for the shape of floating structure, and assessing their dynamic behaviour under the action of waves, winds and currents. Structural design of the hull and platform was carried out to assess hull weights and characteristics Tank testing was carried out on a number of basic hydrodynamic forms and various anchorage configurations have been tested in conjunction with different hull shapes.

    A series of tank tests was carried out in 1978 with the objective of minimising anchorage cable forces, and confirming the behaviour of design derived from computer studies. Experimental results deviated from predicted values by considerable amounts, leading to further tests of models designed to explore various aspects of the behaviour of thethered bouyant platforms under wave action and paying particular attention to the analysis of resonance effects in the anchorage cables.

Title : Mobile Platform for Power Generation Based on the Gas Production of Small Oil Fields	Project N° : 03.32/76
Contractor : DEUTSCHE BABCOCK & WILCOX Address : Postfach 100 347 48 4200 Oberhausen 1 Technical director (or person to contact for further information) : M. Bitterlich	Telephone N° : 0511 19321 Telex : 922851

1. Task

It is aim of this development project to prove the economic and technical viability to utilize marginal gas fields and associates gas for power generating purposes.

2. Borderline conditions

For reasons of economic comparison we have taken the "Bundesrepublik Deutschland" as an significant area of consumption. Price of electric power and primary energy costs in this region are taken as valid for this calculation. Within the framework of this study the conditions of existing marginal fields are investigated and the possible prices for the supply of N.G. are calculated. Operating time on one location in a marginal field with a production of approx. 300 - 600 MW is assumed to be 5 to 7 years.

3. Platform concept

According to the borderline conditions mentioned under para. 2 we have found out, during the concept phase, which ended in April 1977, that a floatable jack-up platform carrying the power station would be the most suitable solution. It is foreseen that the platform will be set up on a lower structure which has specially been designed according to the prevailing local conditions. It was found that the most economic position for the joints should be approx. - 10,0 m below the surface.

- The overall dimensions: 66 x 66 x 24 m
- Load carrying capacity for
  accomodations and equipment: 9,000 to
- Proposed location: Southern and Central North Sea

The special devices of the jack-up system permits to make the set-up during 70% of the prevailing weather conditions during the summer.

4. Power station

During the conception phase we have investigated the following power generating cycles:

- gas turbines only
- combined cycle with waste-heat boiler
- pressurized furnace boiler with gas turbine.

It was found out, that the combined cycle with wasteheat boiler produced the lowest generating costs up to 400 km distance from the coast. This holds true for primary energy prices who are within the production costs and return expectations of the oil companies providing the gas.

5. Energy transfer

For technical reasons, the energy transfer can only be achieved by means of an high voltage direct current system. Only this system permits the transport of electric energy over large distances by sub sea cables. The evaluation of already existing cables placed on the surface of the sea bottom revailed the relative sensitivity against damage of the so placed cables. Subject of the study is also to evaluate existing trenching methods in view of their performance and to propose eventually corrective action.

Title : Production from Arctic Zones	Project N° : 04.04/76
Contractor : GERTH Address : 4, av. de Bois Préau 　　　　　92502 Rueil Malmaison — Paris Technical director (or person to contact for further information) : M. Leblond	Telephone N° : 749.02.14 Telex : 69066F

In each of the areas below the following work has been carried out.

- Data collection.
  Evaluation of the risk of being struck by an iceberg, feasibility of
  removing icebergs by towing. Measurements of the thickness of ice flows.

- Soil investigation.
  To demonstrate the possibility of recognizing the nature of morainic
  utilising various available seismic tools.

- Under-sea civil works.
  Carry out successful tests of a system for under-sea excavation using
  a bucket wheel.

- Under-sea equipment.
  Engineering studies concerning well heads, their cost and operational
  procedures.

- Vertical liaisons.
  Final study of the production riser and definition of the characteristics
  to be retained in the case of the utilization of the platform DYPOSER;
  sliding housing, ball and socket.

Title : Separation Processes	Project N° : 04.08/76
Contractor : BP TRADING LTD. Address : Brittanic House Moor Lane – London EC249 Technical director (or person to contact for further information) : D.K.Knights	Telephone N° :(01)920.8000 Telex : 888811

Brief final review

Work on this contract was aimed at the reduction in size and weight of gas/oil and water/oil separators thus reducing the cost of offshore platforms and consequently offshore production.

The gas/oil separation development has concentrated on the use of a cyclone. A twin cyclone system was developed and tested in Kuwait at throughputs up to 10,000 barrels/day. The results obtained with this unit gave specification gas and oil streams with low additions of antifoam agent (5 ppm).

A system based on the use of cartridge coalescers has been developed for the dewatering of crude oil and subsequent effluent water decoiling. The coalescers are protected by a backflushable prefilter.

The 400 barrel/day pilot plant unit, based in the East Midlands oilfield attained the targets of less than 0.5% water in oil and less than 15 ppm oil in water.

Sufficient information has been gained to enable designs of prototype gas/oil and water/oil units to proceed with confidence.

Title : Improved Crude Oil Production and Treatment	Project N° : 05.01/76
Contractor : BP TRADING LTD. Address : Brittanic House Moor Lane — London EC249 Technical director (or person to contact for further information) : D.K.Knights	Telephone N° :(01)920.8000 Telex : 888811

The overall objective of the project was to increase the amount of usable crude oil produced during primary and subsequent recovery phase in North Sea reservoirs. A range of in-reservoir oil displacement fluids was investigated. In particular a microemulsion fluid was developed to raise the injectivity of water injection wells by displacing flow impeding residual crude oil out of the critical flow zone around the well perforations. This development was continued through rigorous laboratory simulation tests and large scale plant blending to a successful full scale test with a Forties field injection well. Other development of surfactants and temperature sensitive polymer solutions was terminated at the laboratory test phase.

The separation of co-produced water from crude oil is a major operation in the later life of most oilfields. Work on identifying the optimum conditions for achieving this within offshore platform restrictions was carried out. A unique test rig was built and was successfully used to identify optimum conditions and chemical additives for North Sea crude oils. This work culminated with a field test (onshore) at the Forties oil treatment plant which substantiated test rig results. The function of the test rig and associated laboratory test procedures has been explained to many European manufacturers of additives. The problems associated with the production of surfactant and polymer contaminated crude oils that might arise from tertiary recovery operations have been investigated.

The developments listed above were supported by an associated programme of more fundamental research. A paper covering part of this project was presented at the EEC Luxemburg symposium in April 1979.

Title : Pilot Injection of Microemulsion and Polymers in the Chateaurenard Reservoir	Project N° : 05.02/76
Contractor : GERTH	Telephone N° : 7490214
Address : 4, Av. de Bois Préau 92502 Rueil Malmaison – Paris Technical director (or person to contact for further information) : M. Leblond	Telex : 69066F

This pilot injection has been carried out on the reservoir of Chuelles (Chateaurenard concession) in an area of 1 hectar (ha), the configuration employed was an inverse "five pot" with sides of 100 m and with an intermediate observation well between the injection wells and the production well. The reservoir is situated at a depth of 600 m and constituted of a level of non-consolidated sandy clay of 3 m thickness which contains a relatively viscous oil (40 centipoises).

The flood of microemulsion (964 m^3) was injected the 7th of February until the 14th of March 1978, the injection of concentrated polymers was begun on March 20th and continued until the end of the year. At this date, 6958 m^3 of polymers of a cumulative total of 7922 m^3 to date been injected (7922 m^3 is approximately 80% of the pore volume of the area).

An addition to the production of oil was observed from three production wells at the end of March, the percentage of oil in the pilot production wells, equal to 14% before injection attained 40% in May (40% to 60% depending on the well) and remained stable at this value until July, a decline then commenced which reached 32% by December . The oil gained by 31.12.78 represents 1,573 m^3 (the difference between real oil produced and the primary production extrapolated) which is 29% of the oil in place in the area prior to injection.

Title : Development of Heavy Oil Processes	Project N° : 05.03/76
Contractor : WINTERSHALL	Telephone N° : 0561 3011
Address : Postfach 104020 3500 Kassel Technical director (or person to contact for further information) : V. Gojo	Telex : 99632

The development of a process for recovery of heavy oil in West Germany is the subject of this project. The Nordhorn field was discovered in 1942, however oil is not recoverable by conventional means.

In 1975, 5 wells were drilled. They should give a clear picture of the geology and dimensions of the reservoir. Most important they will give rock and oil samples from all wells as all the wells have been completely cored. Very little oil has been recovered from the cores, all the wells were perforated and 300 litres of oil were recovered from 2 wells. This oil had accumulated in the bore hole over a period of some weeks. The laboratory tests carried out were:

  Evaluate the rock physics data
  Evaluate oil properties
  Flooding tests with hot water and vapour
  Tests of in situ combustion
  Solvent tests with organic solvent
  Research into uses of the oil

Because of the high oil viscosity 1,000,000 centipoises at reservoir conditions only thermal methods and the use of solvents seem appropriate. At well Nordhorn 1005 two steam injection tests (Nov. Dec. 1977) had to be abandoned after 4 days because of too small absorbtion by the reservoir. The 2nd injection test (July—August 1978) was applied using higher pressures and the well absorbed within 14 days 1,000 tn of vapour and the production test obtained 56 tns of oil and 290 m^3 of water.

No further field tests at Nordhorn are envisaged until a better understanding of the uses of the oil is obtained, the following possibilities will be looked at:

1) Use as fuel in power stations
2) Prime matter in the manufacture of bitumen
3) Prime matter in coaking fuels.

Tests carried out so far have shown that steam flooding and steam stimulation can be possible recovery methods however it is not clear yet whether the oil can be economically produced.

Title : Research on Improved Hydrocarbon Recovery from Chalk Deposits ("Chap" Project)	Project N° : 05.04/76
Contractor : SHELL, The Hague Address : 30, Carel van Bijlandtlaan Den Haag Technical director (or person to contact for further information) : B.A.Lavers	Telephone N° : (070)776655 Telex : 31005

The "chap" programme as approved by the EEC was initiated mid 1976. It comprises three basically independent lines of investigation, each proceeding stepwise from laboratory studies and experiments to field evaluation of the developed technique, if warranted. The research phase of the project is being finalised and has resulted in the development of:

- an improved productivity evaluation technique, utilizing a special logging tool for fracture detection and for determination of the fracture orientation.

- a drainhole drilling technique employing ultra-high angle wells drilled perpendicular to the least in-situ principle stress.

- a stimulation technique consisting of a viscous fingering acid fracturing technique in which the fracture is subsequently propped with a non-crushing proppant.

Combination of these techniques may reduce costs of development of chalky reservoirs. Attempts to have the techniques field tested have as yet been unsuccessful.

Title : Exploration of Bitumenous Shales	Project N° : 05.05/76
Contractor : GERTH	Telephone N° : 7490214
Address : 4, Av. de Bois Préau 92502 Rueil Malmaison - Paris Technical director (or person to contact for further information) : M. Leblond	Telex : 69066F

The project consists of two principal phases:

- the technology of pyrogenation
- the technology of exploitation in situ.

I. Technology of pyrogenation

Tests were carried out in the laboratory under various atmospheres and with various heating programmes to determine the optimal pyrogeneration conditions for the shale of the Paris Basin. For a series of conditions retained as the most promising pyrogenation of important charges of shale (50 to 100 kg) can be effected leading to the production of quantities for valorization studies.

Pyrolyses tests on a very tall pilot installation were carried out at a contractor leading to the setting out of the necessary elements for a technical and economic evaluation. The LURGI-RUHRGAS process because of its characteristics and its situation in Europe was retained in opposition to the other processes. Complementary tests were carried out in the laboratory to test a variant of the LURGI process. If the results are favourable, LURGI may be asked to modify their installation in this sense.

Concerning the results of the pyrogenate studies one may here and now retain the following indications:

- Highly sulphurated and nitrated shale oil is strongly unsat-
  urated, the yields are higher in middle distillates and reduced in
  heavier distilates

- Concerning the LURGI tests it seems that the yields of oil have been
  less than anticipated.

II. Technology of in situ exploitation

Two processes of in situ shale-treatment have been actually envisaged. In the plain sense of the term, the work will be orientated to surface operations, drilling, creation of fissures by hydraulic fracturation, pyrogenation, recovery of oil.

The process already tested by Occidental Petroleum consists of a previous mining of 15 to 20% of the layer, one carries out an explosion, followed by an underground combustion in the caved in zone of the chamber created by the explosion.

The programme implemented is the following:

- a study of combustion in blocks of shales
- a study of combustion in grains of shales
- a study in a pilot area of combustion and pyrolysis of heavily crushed shale.

These studies have been completed and the following general conclusions advanced:

1) In situ methods are definitely less advanced than ex situ methods.
2) The combustion tests on grains (= 5mm) and blocks (= 250 mm) were of an exploratory nature and it was obviously not possible to determine the optimal aeraulic and thermal conditions for processing this Toarcien oil shale.

Title : Swell Damper	Project N° : 06.05/76

Contractor : BERTIN
 	
Address : Zone Industrielle
       B.P.3 — F 78370 Plaisir
                 France
Technical director (or person to contact
for further information) : M. Facon
	Telephone N° : 462 2500
	
	Telex : 26619

     The study led to the demonstration of the feasibility of a swell damper. In effect, the amplitude attenuation confirms, if not betters, the specification.

     A floating version of the swell damper, less cumbersome than initially concieved, was studied in detail both from a theoretical and experimental point of view and formed part of the pre-project study. This solution presents an additional advantage, it could be used for energy production. In effect, it is possible with the aid of turbines operated by air trapped in the compartments to produce significant amounts of energy. As well as carrying out its role as a swell damper the device can also provide electrical energy. The average power available per module is in the order of 1 megawatt. This advantage constitutes an additional argument for this type of damper because it can be used to supplement the platforms energy sources while it protects the platform from the waves.

Title : Equipment of a Special Ship for Soil Investigation	Project N° : 07.01/76
Contractor : PREUSSAG AND PARTNER	Telephone N° : (0511) 19321
Address : Postfach 4829 Arndtstrasse 1 - 3000 Hannover 1 Technical director (or person to contact for further information) : Dr. Amedick	Telex : 922851

Components of the total system:

- Drilling equipment
- In situ investigating equipment (Penetro meter).
- Geophysical equipment
- Vibrocone equipment
- Ship "Berliner Tor"

Drilling device and coring equipment for unconsolidated sediments were developed. Experience in solid rock could not yet be gained. The penetro-meter has been carefully tested and modified.

The geophysical equipment is a combination of Precision-Echosounder, Side Scan Sonar, Boomer, Sparker and a digital system. The system has been tested and is operational. The development of processing-programmes has been finished.

Two vibration devices ("Kieler Hammer" and "Senkowitch") have been tested on the coring device winch that will be installed and are operational.

The ship "Berliner Tor":
Tests, sea trials and mooring tests have demonstrated that the ship suits very well for use in the southern part of the North Sea.

Title : Soil Investigation in the North Sea	Project N° : 07.02/76
Contractor : FUGRO CESCO	Telephone N° : (070) 209250
Address : Veurse Achterweg 6 Postbus 41 — 2260AA Leidschendam — Holland Technical director (or person to contact for further information) : **A.J.A.** **van Overeen**	Telex : 31010

a. <u>Anchor packer</u>

After the seatrials in May 1978, where the basic feasibility of off-shore use was proved, a lot of additional effort had to be spent on existing drilling and testing system. As usual quite some additional investments had to be done to carry out the necessary field test to prove the technical feasibility and show the marketability to clients. This latter proces is expected to take some two years more before a final evaluation can be made about the success or failure of this part of the research project.

b. <u>Seabed jack</u>

In November 1978 the seabed jack was tested offshore. This first use showed besides some minor technical problems the overall technical viability of the concept.

The technical problems concerned mainly the underwater full opening of the clamp. The tests showed that a lot of simplification in operation and design will be required before the tool can be eventually offered economically. The result of this part of the programme consists mainly of an increased level of knowhow for this type of systems and its impact on standard operations.

c. <u>Re—entry system</u>

This more simple approach to understand problems on rigging and seabed stability had its influence by improving the design of the seabed jack, above.

No action was taken to develop this particular design further for better performance, as it proved to be basically awkward.

d. <u>Conclusions</u>

Overall, in this project a much better understanding was gained on vertical drillstring control. One of the three prototype tools developed showed enough promise to develop further operationally to try to find an economical use for it.

Title : Development of Submersible and their Supply Boats	Project N° : 07.05/76
Contractor : BRUKER PHYSIK	Telephone N° : 0721-5118589
Address : Silberstreifen 7501 Karlsruhe-Forcheim Technical director (or person to contact for further information) : Dr. Ing. Witte	Telex : 826836

The submersible of the mermaid class actually working in the North Sea was permanently improved and adapted to different underwater working jobs. The improvements are partly performed in close cooperation with the operators of the subs.

The manoeverability of the subs have been improved drastically by introducing a hydraulically operated steerable main thruster.

The Bruker Diverlockout Submersibles have been equipped in the mean-time with closed circuit deep diving systems for lockout operations in 200 m water depth and more, with observance of the latest regulations and rules of classification authorities such as Norske Variates, ABS and GERM. LLOYD. tests carried out the last few months proved to be quite successful.

For the mating system between lockout submersible and existing DDC'S flexible spool-prices have been developed by Bruker and fully tested by the contractors.

Another essential component, an energy-saving diver heating system, especially for lockout submersibles, has been developed. This system will double the actual bottom times.

The development of a tool system for a hydraulic manipulator has also been completed. Prototypes of grinders, rotary brushes, cable cutters and wrenches have been tested.

A deep diving submersible with dry transfer, rescue and lockout facilities has been constructed and meets the requirements for more pay-load and energy supply for lockout operations coming from offshore con-tractors.

Development and construction of a diving support catamaran. This project was a little delayed due to the necessity to complete other orders. However the development is now complete and has led to the SUBCAT a vessel of 800 tons, as well as the design of an energy supply buoy crucial to the SUBCAT operation because of the limited possibilities of energy storage on board.

Title :A Submerged Vehicle Tool System	Project N° : 07.08/76
Contractor : WINN TECHNOLOGY	Telephone N° : Bandon 49601
Address : Kilbrittain co. Cork	Telex : 8443
Technical director (or person to contact for further information) :	

1. A claw device has been developed which carries out a number of the requirements without the employment of special tools.

   This device has been fitted to the machine developed under contract
   13/75 and has proved capable and adaptable in many situations.
   It is capable of continuous rotation and can therefore be used to
   operate rotating tools without the addition of an extra power unit.
   A special grip receptacle is provided to enable tools to be gripped
   concentrically and thereby rotated.

2. Screwdriver tools, drills, spanners, and saws can be provided with this
   grip, but it should be noted that on many occasions the claw jaws are
   capable of gripping nuts, bolts, valves etc. without additional tooling.

3. The rotary brush tool is held in the claw device described in 1) above
   and can undertake various tasks through the employment of various sizes
   and grades of bristle. Brush diameters up to 22" have been employed.

4. An ultrasonic surveying head has been developed incorporating a liner
   array of transducers enabling objects to be detected. This can be
   mounted on the rotating portion of the wrist to provide the necessary
   scanning in azimuth and bearing.

5. A submerged stud welder has been demonstrated wherein the energy
   imparted to the weld produces a pressure rise in the molten metal
   exceeding ambient sea pressure, hence producing metallurgically
   uncontaminated weld.

6. An impact driver has been developed: results from this have been disappointing due to limitation of terminal velocity achieved from internal mass. An electro-magnetic device is under investigation, using a
   pulse power supply developed for the welding device described in para
   5) above.

7. Seismic investigation of subsea surfaces has proved possible using the
   impact driver described in 6) above in conjunction with the ultrasonic
   head referred to in para 4).

   Operation of these devices is readily possible in the company's in-house
test tank facility where the necessary operator training has been carried out.

Title : New Technology for Pipelaying at Sea	Project N° : 09.06/76
Contractor : BOUYGUES	Telephone N° : 630 2311
Address : 381, Av. du Général de Gaulle 92140 Clamart – France Technical director (or person to contact for further information) : M. Auperin	Telex : 204420

In response to the evolution of the market in pipelaying we requested in June 1978 a modification of our contract to orientate the study towards laying in deep water with a long, supple stinger.

The areas which were studied in 1978 were as follows:

- the behaviour of laying barges in deep water without the aid of dynamic positioning
- a study of S-curves in deep water
- the definition of a long, supple stinger for a depth of 2,500 m
- a study of the feasibility of a model laying test at a scale of 1/10.

The results obtained on the above programme and the further evolution of the pipelaying market led us to request the termination of our contract in November 1978.

Title : Laying of a 12" flexible at 550m	Project N° : 10.04/76
Contractor : COFLEXIP Address : 23, Av. de Neuilly 75116 Paris Technical director (or person to contact for further information) : J. Laurent	Telephone N° : 7470530 Telex : 18024

   At an earlier stage, we have fabricated a prototype and simulated on
land the working conditions of the flexible.

   − hydrostatic crushing due to external pressure
   − static acial traction
   − dynamic movements

a) To determine the resistance to collapse (55 bar) with the outer layer
   pierced, the following material was utilized.

   − Compression chamber
   − Pressure delivery system
   − Reduction in volume measuring system
   − Measuring thermometre

   Before being put into the chamber the flexible is filled with water and
   connected to the measuring system. While the pressure in the box is increasing, the flexible decreasing in volume expels some of the water
   which it contains. This water collected in a graduated cylinder gives
   the amount of volume decrease.

b) To verify the tensile strength (160 t), the force corresponding to a
   flexible submerged at 550 m, the following material was utilized:

   − a tensile machine
   − measuring apparatus

   The sample was held at one end on a fixed point of the bank by a ferrule,
   the other extremity was attached to the jack of the traction block.

c) A study of the behaviour of a flexible laid on the bottom but not touching
   the soil and therefore exposed to cyclical forces by the currents was
   examined with the aid of a fatigue machine.

   The **bending is given by the inclination of the jaws which hold the extre-**
   mities of the flexible, rotation is implicit in this arrangement.

Title : Offshore LNG Transfer System	Project Nº : 10.06/76
Contractor : DAVID BROWN—VOSPER (OFFSHORE) LTD	Telephone Nº :
Address :	Telex :
Technical director (or person to contact for further information) :	

The objective of the work is to establish the design of an offshore LNG transfer system suitable for duty in the North Sea.

An assessment of the transfer system design requirements has been completed, various mechanical concepts evaluated, a gas—tight LNG cryogenic swivel has been successfully designed, built and tested and a small scale model built and demonstrated.

The next phase is to seek the involvement of one or more major oil companies to participate in the building and testing of large and full scale hardware.

Title : Cryogenic Pipeline	Project N° : 12.03/76
Contractor : O.T.P. Address : 3 & 5 Rue Volta 92 Puteaux — France Technical director (or person to contact for further information) :Nguyen Van Tuyen	Telephone N° : 506 2194 Telex : 62495

The objective of this study was to examine the possibility of transporting LNG long distance. Several aspects have been considered:

- technical and technological problems
- transportation economics
- possibility of application in Community countries
- environmental impact.

Beginning with the existing state of the art and existing materials it is possible to demonstrate the technical feasibility of an LNG pipeline. Several solutions pertaining to the structure of the line were developed.

The economic study, taking into account the actual price of steel or nickle, allowed a comparison between the transportation of natural gas in its gaseous form and its liquid form — in certain cases the transportation of LNG can be more economic — this method of transportation permits the retrieval of refridgeration at the pipeline terminal.

Cryogenic transportation may be the solution to moving large quantities from producers to consumers. Certain schemes may be found in the Community and in other countries of the world.

1977

3rd Round Projects

Title : Marine Seismic Source Development	Project N° : 01.05/77
Contractor : HORIZON EXPLORATION LTD.	Telephone N° : Swanley 68011
Address : S&A House – Azalea Drive Swanley – Kent BR8 8JR – UK Technical director (or person to contact for further information) : M. Greener	Telex : 896050

During 1977 and early 1978 Horizon Exploration Ltd. (formerly S & E Geophysical Ltd.) carried out a project to evaluate the effectiveness of a new marine seismic energy source. The new source known as a "water gun" was compared with the results obtained from the conventional "air gun" energy source.

The principal objective of the project was to compare the two sources in a variety of geological prospects. It was estimated that the project would take 12 months and cost pounds 201,500.

In the event the programme had to be modified slightly owing to bad weather conditions, part of the series of tests being abandoned and other aspects being extended. Nevertheless the main objectives were 90% achieved. The project took 13 months and cost pounds 156,051, of which the EEC provided a loan of £ 62,420.

The results showed that in certain geological prospects the characteristics of water guns would lead to better seismic data than air guns. The detailed results have been the subject of papers to international meetings of exploration geophysicists and to the EEC sponsored Symposium in 1979.

The project has now reached the point of commercial exploitation with several seismic surveys using water guns which have been carried out in the North Sea for several major oil companies in the summer of 1979. The results of these surveys having shown a marked improvement on previous seismic data.

Title : Seismic Prospection Using Transverse waves	Project N° : 01.09/77
Contractor : C.G.G. Address : 6, rue Galvani B.P. 56 - 91.301 Massy Technical director (or person to contact for further information) :M.Mennebeuf	Telephone N° : 9208408 Telex : 692442

1. The project consists of three phases:

   - a phase 1, comparing sources of transverse waves
   - a phase 2, testing various methods
   - a phase 3, of definition and setting up methods of interpretation.

2. Phase 1

   Those sources have been tested on an experimental profils situated in the forest of Orleans. - an explosive source. Practically a string of explosives was sunk in those parallel sections. The explosion in the central range creates a dissymmetry in the middle. The explosions in the lateral ranges furnish an energy P of the polarised cutting energy S in the privileged directions opposed to the zones of dissymmetry. The P waves are almost eliminated by subtraction of the lateral recordings.

   - two mechanical systems, the marthor developed by IFP and which delivery horizontal impacts transmitted to the soil from an anchored target, secondly a horizontal vibrator where the vibrating plate is attached to the soil by means of teeth.

   The recordings obtained were compared and the three sources may be considered equivalent from the point of view of the quality, the consideration of choice is practicality, humid terrain is unfavourable for explosives, access problems and damage of passage affect the mechanical systems.

3. Phase 2 and 3

   The recorded data of the S waves have been tested by conventional methods without problems. The classical techniques of speed analyses have been equally applied with success.

   The work has therefore focused on the combination of results in mode P and S by association of the P and S horizons corresponding to the same geological discontinuities and on the calculation of the gamma coefficient (return of the internal speeds between the associated horizons and established by making the return times of the distance travelled independent of the mode P and S).

One is finally left with a fine correlation, that is to say, methods to associate the P and S times where the seismic correlation of the two modes is bad.

Two programmes have been written and several theoretical studies made.

Title : Deep—Sea Drilling Techniques	Project N° : 02.09/77
Contractor : GERTH	Telephone N° : 749.02.14
Address : 4, Av. de Bois Préau 92502 Rueil Malmaison — Paris Technical director (or person to contact for further information) : M. Leblond	Telex : 69066F

The discovery of hydrocarbons in ever—increasing depths of water and in increasingly difficult seas has resulted in changing methods of drilling. Dynamically-positioned ships were build in the early 1970's enabling drilling operations to be carried out from independent vessels. Application of such techniques underscores the governing points of their optimum use in increasing depths of water, and these points are the purpose of the studies conducted under this contract.

Dynamic positioning

Integration of a DOPPLER SONAR into the dynamic positioning system, together with improvement to the long—base acoustic positioning system gave rise to successful trials in November 1979.

Riser

The riser enables the drilling rods to be isolated from the sea bottom up to the surface. Work carried out in 1980 covered the four main factors that are essential to future drilling operations in very deep waters, namely:

— Study of the behaviour of the riser under the action of the longitudinal vibrations encountered at depths of over 1500 m,

— Architecture of the riser and its tensioning devices,

— Lowering weight of the riser by using suitable buoyancy materials and building it in new materials such as titanium; the fatigue tests on a prototype titatium riser will be continued in 1981,

— Improvement to the riser foot pup joint, by adding a rod joint sensor and a gas detector.

Re—entry sonar

The re—entry sonar makes it possible to seek out the wellhead on the sea bottom when the ship has no method of accurate location at the surface. A re—entry sonar has been developed and demonstration trials for this sonar performed.

3000 m drilling support

Study of a conventional vessel capable of making drillings in depth of water of 3000 metres was completed in 1980. This vessel will be equipped with dynamic positioning and a derrick with a hoisting capacity of 900 tons, supplemented by tensioning devices and a heave compensator. The study file drawn up will provide a starting point for study of the naval architecture that is indispensable to the construction of such a vessel.

Title : Seabed Hydrocarbon Production	Project N° : 03.33/77
Contractor : DEEP SEA PRODUCTION SYSTEMS	Telephone N° : 01.837.3377
Address : 40, Bernard Street London WCLN 1LG Technical director (or person to contact for further information) :M. J. Collard	Telex :22308

The project seeks to develop the principle of dry one-atmosphere containment of production equipment on the seabed for the purpose of exploiting marginal and deepwater reserves. The programme has been divided into two phases:

Phase I — Technical & Economic Feasibility of Concept
           March 1977 to June 1979

Phase II — Development of Selected Designs
           July 1979 to November 1982

The long term objective of the project is to provide an economic means of developing isolated deepwater fields: so far water depth of between 500 and 1000 metres have been considered. A number of inter-connected reinforced concrete cylinders on the seabed contain manifolding, separation, water injection and perhaps gas compression equipment in a dry one-atmosphere environment. They are connected to the surface for power generation, utilities, tanker loading and gas flaring. Control and maintenance is carried out by crews transported from the surface by sub-mersible or other means.

In addition to this long term deepwater application, a satellite manifold chamber is being developed for shallower waters. It serves marginal satellite fields by collecting produced crude from a number of wellheads and commingles into a larger flowline for transport to a distant processing platform for treatment. The manifold is normally unmanned and access is necessary only for periodic inspection and maintenance.

In developing the overall design concepts several particular technic-al areas have been itemized for detailed development and testing. These include the reinforced concrete and composite pressure hulls, the method of access to the seabed facility, the surface support structure, the installation on the seabed of the chambers and the development of flex-ible underwater power and control cables.

66

Title : N° 2 Production Pilot Project in 100–200 M of Water	Project N° : 03.34/77
Contractor : GERTH	Telephone N° : 749.02.14
Address : 4, Av. du Bois Préau 92502 Rueil Malmaison – Paris Technical director (or person to contact for further information) : M. Leblond	Telex : 69066F

In order to qualify the relevant technology, the purpose of the project developed under this contract was to test a complete deep water offshore production system under actual operating conditions that was to be installed on ALWYN field, situated in block 3/9 East of the British zone of the North Sea.

The refusal by the authorities to allow the associated gas of ALWYN field to be burnt forced the contracting party to delay development of this field, thus interrupting the associated pilot research project at the study stage.

An alternative solution to continue the present contract has been sought on the FRIGG North-East gas field, though resumption of the pilot project on this field was dependent on the agreement of the co-holders of the permit.

Since this agreement was not forthcoming, the work was abandoned on 30.06.80.

The work performed at ALWYN was confined to study of the overall production scheme and the critical elements.

The work on the subsea well base studied under this contract will be used for the development of FRIGG N.E. field.

Title : Deep Sea Production Equipment Techniques (T.E.P.M.P.)	Project N° : 03.35/77
Contractor : GERTH	Telephone N° :749.02.14
Address : 4, Av. de Bois Préau 92502 Rueil Malmaison — Paris Technical director (or person to contact for further information) : M. Leblond	Telex : 69066F

This project involves study of the basic components that are implicit in any general offshore petroleum exploitation project in deep waters involving subsea wellheads, a floating production support, links between the bottom and the surface and appropriate product evacuation facilities.

The purpose of the project is to obtain the necessary engineering data, to build prototypes and to test the critical components so as to define their limits of use.

The work covered the following components and processes:

Anchorage: A prototype detector of anchor line faults was tested in the laboratory on samples of anchor cables featuring significant defects.

Riser: The hydrodynamic loads that a production riser undergoes were subjected to systematic tests on a testing tank model. The results of these tests have been analysed and the general architecture of the riser defined.

Subsurface link: While not calling the principle of the method into question, experiments conducted in 1978 on a flexible arch in the Mediterranean were a failure, owing to the inadequacy of the logistic facilities and sea states that were abnormally poor for the season.

Laying and connection of a flowline: The flowline laying and connecting method studied under this contract is based on a J-configuration laying method; the method did not qualify for a demonstration test, since use of surface supports would have resulted in very high costs. A test based on a method of towing near the bottom was preferred to it and is to take place during the first half of 1981.

Operations inside wells using the T.F.L. technique: Adaptation of the T.F.L. technique to lines with a diameter of 4" was successfully achieved, while at the same time enabling the endurance of the equipment to be analysed and the equipment best suited to pumped-tool operations to be selected.

Electrical connectors that can be plugged-in underwater: Testing of DEUTSCH connectors was continued throughout 1980 to simulate several years of underwater operation.

Horizontal diphasic flows: Laminated flow and plug flow conditions were tested in 6" loop line. Test equipment and multimetric models were developed to enable systematic experiments to be performed.

Pollution prevention: A simulated programme of rising of oil and gas through water to the surface was tested and technical manuals on using the design programme were drafted.

Title : Maintenance of Subsea Equipment. Experimental Programmes at Grondin N.E. Station	Project N° : 03.37/77
Contractor : GERTH Address : 4, Av. de Bois Préau 92502 Rueil Malmaison — Paris Technical director (or person to contact for further information) : M. Leblond	Telephone N° : 749.02.14 Telex : 69066F

Within a few years, offshore oil will be being exploited in depths of several hundred metres of water, where human intervention by divers is no longer possible.

When this time comes, methods of working both inside and outside wells will hence be needed (using robots or submarines) to ensure that submerged production equipment can be correctly maintained.

The purpose of the present contract is to try out these operating means, firstly using the testing facility formed by the GRONDIN N.E. experimental subsea station, and secondly a rail-mounted subsea remote-handling device built for the purpose as part of the COMEX-SEAL 6/75 contract.

Operating at the bottom of wells was successfully tested using devices enabling the work to be performed despite the heave movements of the support vessel.

Operations outside the wells from a manned submarine equipped with special tools showed that the well can be approached "blind" and the valves and safety equipment of the GRONDIN N.E. station installations can be actuated.

Construction of a remote-handling device robot known as the TIM (French acronym for remote-handling and operating device) has been completed. This device is equipped with two hydraulic arms capable of handling a weight of a hundred kilos with a precision of one centimetre. A crane placed at the centre of the device also enables a weight of one ton to be lifted. Tank tests were conducted before starting a campaign of operating and handling trials at GRONDIN N.E. station in the first half of 1981.

Title : TFL Techniques	Project N⁰ : 03.39/77
Contractor : THE BRITISH PETROLEUM COMPANY LIMITED Address : Brittanic House Moor Lane — London EC24 9BU Technical director (or person to contact for further information) : P. Smith	Telephone N⁰ : 628.4090 Telex : 888811

Through Flow Line (TFL) servicing of wells, initially introduced in the Mexican Gulf during the 1960's was confined to scraping paraffin wax from the tubing of highly deviated platform wells, which was difficult to perform by conventional wire line techniques. Initial TFL success led to the carrying out of a variety of downhole operations.

With the development of fields such as Buchan and Magnus with their subsea satellite wells and facilities such as the tethered buoyant platform, BP decided in 1977 to embark on an in-house evaluation of existing TFL technology.

The project was split into two Phases. EEC Contract N⁰ 03.03/77 only provided support for Phase I.

Phase I of the project which was successfully completed early 1980 consisted of the following:

1. An assessment of the performance of a 3" TFL system in a short surface loop at Montrose, Scotland.

2. The development of a method of producing "TFL Quality" welds in offshore flowlines which could readily be applied to the new reel-barge laying methods.

3. The construction of a typical 4" nominal subsea satellite completion at BP's oilfield at Eakring, Nottinghamshire, comprising a 2000 m surface loop and a 440 m well, and an investigation of wear, control effects and tool reliability when pumped over long distances for a considerable period.

Title : Novel Gas/Oil and Water/Oil Separators	Project N° : 03.40/77
Contractor : BRITISH PETROLEUM COMPANY LIMITED	Telephone N° :(01)9208000
Address : Brittanic House Moor Lane — London EC24 9BU	Telex : 888811
Technical director (or person to contact for further information) : D. Knights	

1. Development work and field trials of novel gas/oil and water/oil separators under EEC Contract N° 03.40/77 have been successfully completed.

2. The new water/oil separator uses filter cartridges to coalesce the water droplets in the feed and thus to enhance gravity separation. In comparison with conventional water/oil separators, the new unit has the following advantages:

    a) It occupies only 20 per cent of the space required for a conventional separator;

    b) Its weight is only 25 per cent of that of a conventional separator;

    c) Despite the savings in space and weight, its performance is superior to that of conventional separators.

    Cartridge life is greater than 50 days which exceeds the original specification by 67 per cent.

3. The new gas/oil separator uses centrifugal force in a specially adapted cyclone to break the foam and separate the gas from the oil. Compared with conventional separators it offers the following advantages:

    a) It requires less than 25 per cent of the deck area;

    b) Its weight is less than 30 per cent of that of conventional separators;

    c) It meets target specification of less than 5 per cent volume of gas in the separated oil with no oil carry-over in the gas. Performance monitoring has demonstrated that such an efficiency could not be achieved with a conventional separator.

These advantages of the new water/oil and gas/oil separators make them particularly attractive for offshore installations.

4. An agreement has been reached with a manufacturer to produce and market these compact separators. First orders are already being progressed.

Title : Water—Oil Separation by High Speed Centrifuge	Project N° : 03.41/77
Contractor : BERTIN & CIE	Telephone N° :056—2500
Address : B.P. 3 — Zone Industrielle F — 78370 Plaiser — France	Telex : 692 471
Technical director (or person to contact for further information) : M. Facon	

A study of the efficiency of removing oil from water, having characteristics similar to water produced with oil, was carried out by SNEA (P) for Bertin under the following conditions:

- tests were carried out in the oil removal test station of Micoulau of SNEA (P) at Pau

- SNEA (P) made up emulsions of oil in water

- the parameters studied were the following:

    a) volume of water and treatment time
    b) nature of the crude and its percentage of the total volume
    c) influence of chemical additives
    d) speed of rotation of the centrifuge.

These tests were carried out in order to determine the influence of the type of crude and the chemical additives on the efficiency of separation of the demulsifier. To complete these tests three **crudes of totally** different chemical and physical characteristics were used, these were:

- Ashtart which is representative of the majority of crudes
- Frigg is a very fluid condensate at ordinary temperatures
- Rispo Mare is a very viscous oil which contains a high percentage of asphaltines which favours stable emulsions.

The chemical additives chosen from the range of chemicals normally used in the field:

- corrosion inhibitors : Norust 720, CK 337
- bacteriocide   : Bactiram
- demulsifier   : d 13 E — Simulsol 2777

The following conclusions could be drawn:

1) Independant of the nature of the crude, for quantities of oil in water of 200 to 2000 mg/l the centrifuge produces an effluent containing less than 20 mg/l of hydrocarbons.

2) The additives do not greatly alter the efficiency except in the case of heavy oils.

3) The centrifuge, as presently concieved, is not suitable for handling entrained solids which are normally present in water produced with crude.

4) The instrument is very sophisticated and may require marinisation.

Title : Extension of Applications for Articulated Columns	Project N° : 03.42/77
Contractor : GERTH	Telephone N° : 749.02.14
Address : 4, Av. de Bois Préau 92502 Rueil Malmaison - Paris Technical director (or person to contact for further information) : M. Leblond	Telex : 69066F

The initial applications in the North Sea for articulated columns were confined to simple functions such as the flare stack and loading station, for depths of water of less than 150 metres.

The work covered by the present contract consists in broadening the field of application of the articulated column concept, in particular to deep waters.

The permanent coupling between an articulated column and a floating production support has been tested in the testing tank, followed by drafting of engineering files to prepare industrial applications either in the Mediterranean with a ship, or the North Sea with a semisubmersible.

At the same time, a number of components specific to articulated columns (rotary joint, multi-passage device, inflecting riser, materials for universal joint, connector for flexible line, etc..) were defined.

The concept of columns comprising several articulations was developed by calculation programmes and testing tank trials.

Title : T.L.P. Tests	Project N° : 03.43/77
Contractor : SCOT LITHGOW	Telephone N° : 0475.42101
Address : Kingston Shipbuilding Yard Port Glasgow — Renfrewshire PA 14 5DR Technical director (or person to contact for further information) : J.Anderson	Telex : 77192

The project follows on from an earlier project carried out in 1975 on the one third scale model T.L.P. which proved that the T.L.P. was a viable concept and that its response in a seaway could be predicted with reasonable accuracy using a linearized frequency domain computer programme. In this earlier project only a single riser was tested.

The 1979 sea test programme had as one of its primary goals, the evaluation of the performance of the T.L.P. and a bank of twelve risers in water depths equivalent to 600 ft.

This goal was essentially attained with the successful testing for 95 days of the platform, the risers and auxiliary systems. The tests simulate, by a Froude scaling ratio of 1 to 3, the operation of a T.L.P. and riser system at a depth of 600 ft when conducted in 200 ft water depth.

During the programme waves equivalent to 50 ft passed through the structure, with no platform or riser malfunction. Particularly noteworthy was the fact that the risers were run successfully during some of the worst weather and sea conditions.

The instrumentation and data acquisition system performed well without major problems, and a considerable amount of good, meaningful and viable data was recorded.

In addition to the structure, riser, tension member and anchorage data obtained a great deal of practical knowledge was also gained for example:

a) the towing of the platform to and from the test site location, with the anchors secured to the bottom of the columns, demonstrated the stability of the T.L.P. under tow at a towing speed significantly faster than predicted

b) The successful deployment of the anchors from the T.L.P. by the T.L.P.

c) The landing and retrieving of the sea-bed template from the platform confirmed the ability to control the template in a level attitude with a level indicator and subsea T.V.

d) Riser running, latching to mandrel on template and tensioning were successfully accomplished despite tidal current with the aid of T.V. monitoring at the template

e) Deployment of a composite riser bundle with little difficulty demonstrating the ease of operation from a structure not subject to heave.

The analysis of the data gathered showed that the frequency domain programme made successful predictions for motion and leg tension response both at the design tension member stiffness and at a reduced stiffness value.

The time domain programme also proved to be successful in predicting the platform motion when pulled off location and suddenly released.

The riser response was generally predicted well by the computer programme in spite of the difficulties encountered in modelling exactly the end rotational stiffeners.

The overall conclusion is that the test programme achieved its object and provided confirmation of the effectiveness of the design tools for use on a tension leg platform and associated equipment.

Title : COSMAR	Project N° : 03.45/77
Contractor : DYCKERHOFF & WIDMANN AG	Telephone N° : 089–92551
Address : Postfach 810280 D – 8000 München Technical director (or person to contact for further information) :Dr.Finsterwalder	Telex : 05–23036

Phase I of the international joint project COSMAR was terminated on schedule in April 1981 after three years. COSMAR stands for "Concrete Structures for Marine Production, Storage and Transport of Hydrocarbons". The aim of the project was to establish new fundamentals necessary to judge the loading and safety of offshore structures. One part of the project was concerned with offshore structures. One part of the project was concerned with extreme environmental conditions such as great water depths or ice loading, as may be found in arctic regions. Another part of the project examined the action of the cryogenic stored goods (up to −200°C) on the materials concrete and steel, and on the structure as a whole.

The choice of specific materials and the development of suitable concretes, structure types and calculation methods served to establish fundamentals for designing and constructing concrete structures offering the same degree of safety and having the same working life as onshore structures.

The subdivision of the project gives a good survey of the priorities in this research programme:

Part project 1: Investigation of Offshore Concrete Structures with respect to static strength.

Part project 2: Investigation of Offshore Concrete Structures with respect to fatigue strength.

Part project 3: Development of Concrete Structures for storage and transportation of LNG and oil including thermal effects.

Part project 4: Offshore Concrete Structures subject to ice and arctic conditions.

Due to the most interesting and unexpected results, it was of crucial importance to continue the project in order to solve the encountered problems and resume the tests.

Phase II will be terminated by the end of 1982 and has a budget of DM 2.5 Mio compared to DM 8 Mio for Phase I.

Title : Riser Pipeline Installation on Existing Gravity Platforms	Project Nº : 03.46/77
Contractor : SHELL U.K. LTD Address : P.O.Box 148 Shell — Mex House Strand — London WC 2R ODX Technical director (or person to contact for further information) : J.B.Cook	Telephone Nº :01.934.1234 Telex : 22585

This report presents the results of a study which was initiated to investigate the feasibility of installing additional pipeline risers on an offshore concrete gravity platform subsequent to the installation of the platform on location. Pipeline risers form the connections between seabed pipelines and deck mounted production facilities and are essential links in the transportation, to and from the platform, of oil, gas and injection water. Whereas the installation of such risers on a steel piled structure is known technology, this is not the case for a concrete gravity platform.

The requirement for installing additional risers could be due to several reasons such as damage to existing risers, insufficient numbers of risers pre-installed to meet field development plans or simply that the existing risers are inadequate for amended service demands in terms of diameter and pressure rating. In view of the large number of concrete gravity platforms installed in the Northern North Sea and the likelihood that one of the above reasons will necessitate the future installation of a riser on such a platform, it was decided to undertake a feasibility study to examine the problem from the two separate view points described below:

a) To study the possibility of fixing pipeline risers between 8" and 36" diameter directly to an existing concrete gravity structure installed in 150 metres of water depth. Special clamps will have to be designed which can be affixed to the structure and into which the pipeline riser can be introduced and securely fastened. Stresses in the pipeline riser can arise from wave effects, current effects, corrosion, internal temperature and other causes and therefore the strength of the system must be carefully analysed.

b) To study and compare the costs of installing pipeline risers between 8" and 36" diameter by means of a separate self supporting tower structure adjacent to the existing concrete gravity structure. This self supporting tower might be of the articulated type and a flexible or sliding bridge type of connection could be the carrier for the connection of the pipeline riser to the main structure.

Conclusions

In preparing the preliminary designs, the Brent D Condeep platform was chosen as a basis for the exercise. The cost of the attached tower scheme is estimated at £22,700,000 and the cost of the separate tower scheme at £12,200,000 each scheme taking approximately 22 months from the point of design initiation to the completion of installation.

In addition to being the cheaper solution, the installation of the separate tower scheme is also far less susceptible to the effects of adverse weather conditions. It is therefore recommended that this scheme be adopted for the installation of additional risers to Condeep type concrete gravity platforms.

Although the Sea Tank and Andoc type platforms have not been considered in the same detail, the cost difference between the two schemes and the operational restrictions associated with the attached tower scheme are such that the separate tower scheme is considered the most appropriate solution irrespective of platform type. Under certain circumstances, however, it may be possible to use external conductor guides for the support of risers which may lead to a cheaper solution than has been studied in the report.

Title : Watertight Shelters for Subsea Wellheads	Project N° : 03.47/77
Contractor : SEA TANK CO. Address : Immeuble IENA 12, rue Le Corbusier 94588 Rungis – France Technical director (or person to contact for further information) : M. Gerbault	Telephone N° : 687.2332 Telex : 200939

1. General Conception

Oil companies operating in offshore fields show an increasing interest in subsea completions. These indeed present many advantages:

- for great depths, the cost of fixed platforms is prohibitive and the installation of wellheads on the seabed tend to reduce the capital invested.

- for medium depths, marginal fields do not justify a high investment, and the subsea completion presents a more economical solution.

- subsea wellheads can complete the equipment of a field by draining peripheral zones.

2. Description

The project consists of series of separate cells (one for each wellhead) connected to a central corridor which is used for access by personnel and for the housing of:

- the manifold,

- safety devices for self–survival in case of accidental cutting off from the outside,

- and possibly, a manifold for pumped down tools, allowing a reduction in the number of pipes to the exploitation platform.

Each cell can be independently filled (for installation of wellheads) or emptied and rendered to atmospheric pressure when necessary.

Conclusion

Actual status of the study shows that concrete shelters for subsea wellheads are feasible.

Main advantages can be summarized as follows:

Ensure a thorough protection of several wellheads sheltered together inside the same structure and operated from a semi-submersible platform. The structure is placed before drilling and is used as a drilling template.

Allow the wellheads to be dried out and brought back to an atmospheric pressure for local inspection and maintenance with free access for personnel with no special diver's training. The possibility of standard wellhead replacement should be retained.

Consequently each wellhead is enclosed in a watertight cell which may be independently either dried out and brought back to an atmospheric pressure, or flooded and opened to the sea.

Adjoin further subsea exploitation equipment if technically profitable.

Provide supplies of energy and atmosphere from the surface, as well as to ensure a maximum level of security for the maintenance personnel with safety devices for self-survival and self-evacuation.

However, reliability and safety remain as major problems. Particular care shall be reserved to them.

In this matter, sophisticated technical solutions shall be developed. These solutions exist and could be adapted.

The main problem is of economical order.

Actually, the proposed system although presenting qualities and technical advantages is an expensive system due to the technology required. Moreover this system applies to ambitious or at least important oil field development projects.

But presently, oil companies try to develop as quickly as possible and for the lowest cost, small fields or even marginal fields located in shallow waters.

Deep sea programmes are presently slowed down by oil companies for economical reasons.

Therefore, it does not appear reasonable for the time being to mobilize great engineering means for this study which could only be finalized in a long run.

Title : Floating Natural Gas Liquefaction Plant for Offshore Liquefection and Loading of Associated Gas	Project N° : 03.48/77
Contractor : PREUSSAG/AKER/LINDE	Telephone N° : 0511-19321
Address : Postfach 4827 Arndtstrasse 1 D – 3000 Hannover 1 Technical director (or person to contact for further information) : D. Meyer Detring	Telex : 922851

A floating natural gas liquefaction plant for the exploitation of marginal offshore gas deposits has been designed, calculated and partly model tested.

During the design process the concept was discussed with major oil companies and modified according to their suggestions.

Questions of safety of men and environment have been solved in close cooperation with classification societies and safety authorities.

The first phase of an Approval in Principle was carried out by the classification society Det Norske Veritas.

The system is characterized by following main components:

- Natural gas liquefaction plant, mounted on a floating tension leg anchored platform,

- floating LNG storage and loading platform,

- transfersystem for LNG and return gas between process platform and storage platform,

- feed gas pipeline connecting the LNG process plant to the central production platform

Main technical data:

LNG production capacity	$385{,}550 \text{ m}^3/\text{h (Vn)}$
LNG storage capacity	$65{,}000 \text{ m}^3$
Loading rate of LNG loading system	$6{,}250 \text{ m}^3/\text{h}$

Main advantages of the system:

- compared with conventional fixed platform/pipeline concepts, building cost is lower and almost independant of the water-depth.

- no foundation problems,

- excellent motion characteristics guaranteeing a minimum of non-operating time even under northern North Sea conditions,

- the system is mobile and can exploit successively several smaller fields,

- the floating LNG plant can be constructed and equipped completely in industrialized areas and would then be towed to location, thus saving time and money for difficult and expensive offshore work.

Title : Development of an Offshore Gas Gathering System	Project N° : 03.49/77
Contractor : DAVID BROWN VOSPER	Telephone N° : 83331
Address : Graphic House Castle St. – Porchester Technical director (or person to contact for further information) : E.M. Roger-Smith	Telex : 96100

    Two types of floating process terminal have been considered, a multi-leg semi-submersible unit (MLSS) and a barge like moored process and storage vessel (MPSV). The MPSV lends itself to construction in a conventional shipyard, and a steel structure is favoured. The MLSS, however, may be constructed of steel or concrete, each material having particular advantages. At the initial stage, therefore, an investigation of both construction methods was made for the semi-submersible unit so that the design offering greatest promise could be chosen for detail design.

    An investigation of the various natural gas liquefaction processes has been made in this study, enabling a simply operated and safe process to be selected.

    Initial investigation of designs for plant capacity at each end of the scale considered showed that an MPSV having satisfactory seagoing performance would nominally be oversize for a 3.4. MMSCMD plant. The MPSV was found more suitable for larger process volumes and so was chosen for the technical part of the study at a throughput of 10.2 MMSCMD.

    The MLSS was found to be very sensitive to topside payload, and so to minimise the gap between state of art designs and that for this study, the production throughput of 3.4. MMSCMD was chosen for this configuration.

    A representative site in the northern North Sea has been chosen in order to define environmental conditions for either liquefaction terminal. Following definition of the design requirements for each terminal these have been used to develop both schemes in sufficient detail for accurate model scale testing and realistic economic analyses to be carried out. Careful consideration of the requirements of statutory bodies for vessel certification has been taken throughout this study, since their requirements have a significant effect upon terminal general arrangements, and every system fitted.

In addition to testing physical models of both the MLSS and MPSV terminals an analogue computer model has been developed to initially parallel the model tests and subsequently extend consideration of environmental conditions to typical storm rotations of wind and sea.

In the economic assessment of both schemes, operating and capital costs have been estimated. Once overall cost for each scheme has been established and system availability calculated and economic analysis may be made, determining potential profitability and the sensitivity of profitability to changes in various parameters such as: field life, inflation, amortisation period, etc.

Title : Improvements to Gas Detection Systems	Project N° : 03.50/77
Contractor : J & S SIEGER LTD. Address : 31 Nuffield Estate Poole, Dorset BH 17 7RZ Technical director (or person to contact for further information) : D.J.HUCKNALL	Telephone N° : 02013-6161 Telex : 41138

Apart from the sudden deterioration in performance of gas detectors due to the presence of poisons, this investigation has shown that many problems may arise due to the construction and manufacture of such devices.

In practice, catalytic sensors lose activity after prolonged use at high temperatures or after operation in gas-rich atmospheres and this may be due to agglomeration of the crystallites of the active component or to breakdown of the refractory support caused either by its inherent chemical instability or by the effects of thermal shock. The latter mode of deactivation is helped by the fact that, using conventional methods of deposition of the refractory metal oxide which involve the thermal decomposition of a water-soluble precursor of that oxide, a poor bond is formed between the metal base and the oxide. It is possible to manufacture a gas detector which displays suitable mechanical strength and resistance to thermal effects by means of the chemical vapour deposition of a suitable oxide. An application for a patent for such a device was filed in the United Kingdom by J. and S. Sieger Ltd. , in December 1977.

As a result of the present work, it appeared that, although little could be done to prevent the initial interaction of the inhibitor and the catalyst, certain catalyst/support systems recovered very much more rapidly than others from this interaction. One such system was based on tin (IV) oxide.

In an attempt to exploit the ability of tin (IV) oxide to recover from exposure to poisons and incorporate the stability of vapour deposited oxides, preliminary experiments were carried out with a palladium/tin oxide sensor produced by C.V.D. The device was of similar construction to a pellistor detector and consisted of a helical platinum coil embedded in a bead of tin (IV) oxide. The bead was produced by the vapour phase oxidation of tin (IV) chloride on the heated metal coil and the palladium was applied from a solution containing a suitable salt.

Results showed that tin (IV) oxide was readily deposited on the coils to produce a thick monolithic coating. After coating with palladium, the resulting bead readily oxidised carbon monoxide and hydrogen and, to a lesser degree, butane. Little catalytic activity was shown in the presence of methane and further work was clearly necessary before a device suitable for trials in an industrial environment could be developed.

Title : Development of Brine Soluble Polymers and Associated Chemicals for the Enhanced Recovery of Petroleum	Project N° : 05.06/77
Contractor :THE BRITISH PETROLEUM CO.LTD.	Telephone N° :(01)920.8000
Address : Brittanic House Moor Lane — London EC24 9BU	Telex : 888811
Technical director (or person to contact for further information) : D. Knights	

The central theme of the project has been to develop polymers for modifying oil displacement and hence improving sweep efficiency, during a water flood operation. Polymers have been assessed against the conditions found in the major European oilfields; that is they must give stable viscous solutions in high salinity brines over a long period of time at high temperatures. A secondary objective has been to develop chemical additives (such as surfactants) which can, by improving oil displacement efficiency, improve the performance of a polymer flood. The problems of maintaining interfacial activity in high salinity brines has delayed obtaining suitable surfactants. Improved surfactants are becoming available but they have come too late to allow detailed characterisation of their influence on polymer properties.

Of the polymers, only poly (saccharides) produced by fermentation processes have proved of any interest. Of these only the scleroglucan biopolymers Ceca CS11 (liquid grade) and L21 have been able to combine adequate viscosity with long term thermal stability in sea water at 90°C. A major problem still remains of the need to purify the biopolymers after fermentation to remove biological cell wall debris.

Title : Pilot Project for Recovery of Heavy Oil by Injecting Steam into the Upper Lacq Field	Project N° : 05.07/77
Contractor : GERTH	Telephone N° : 749.02.14
Address : 4, Av. de Bois Préau 92502 Rueil Malmaison — Paris Technical director (or person to contact for further information) : M. Leblond	Telex : 69066

Whereas steam injection into sandy reservoirs is a proven and industrially employed technique to recover viscous oils, this type of enhanced recovery has never been attempted on limestone reservoirs and in particular cracked or fractured limestone reservoirs. The purpose of the present contract is to test continuous injection of steam into a cracked limestone deposit so as to evaluate the efficiency of the method which, should it prove positive, could have a very broad field of application, in particular in European countries.

With this in view, a pilot project for continuously injecting steam into the UPPER LACQ heavy oil field was set up. Following two years of injection at a uniform rate of 160 tons per day, the conclusion is that the fissurations in the deposit have in no way hindered the action of the process.

The various parameters continuously followed (bottom temperature, advance of condensed steam by measuring the salinity of the bottom waters and tracing the steam injected)have shown that owing to the high density fissuration, the field behaves, overall, homogeneously.

Interpretation of the results has shown that location of the pilot project at the summit of the structure is a favorable factor: the oil still present in the cracks at the top of the structure was driven by the condensed steam (top water drive) and this oil then joined the output resulting from thermal drive. Under these conditions, those sections of the deposit outside the summit do not profit from this mechanical drive effect: extension of the steam—drive to the scale of the UPPER LACQ field appears to be limited to a few summit cracked zones.

This drive on the pilot block will probably require a number of years of injection before it can be decided whether to extend the method to the scale of the UPPER LACQ field.

Title : Use of EOR Processes in the Cortemaggiore Field, Italy	Project N° : 05.08/77
Contractor : AGIP S.p.A. Address : 20097 S. Donato Milanese 20100 Milano — Italy Technical director (or person to contact for further information) :Prof.Ciarichi	Telephone N° : 53531 Telex : 31246

The purpose of this project is to evaluate different EOR processes potentially applicable to the Cortemaggiore Field, Italy, and to test one of such processes in a field pilot. Cortemaggiore is a conventional light oil reservoir, exploited since 1949; all the wells existing at the beginning of the project were exhausted.

A new reservoir study, carried out using the most up-to-date reservoir geology and engineering methods, brought to evidence a new configuration of the oil-bearing levels, and evidenced their partitioning by tear faults. A numerical model study of the five most promising pools evidenced the presence of areas of the field not completely swept by the natural depletion mechanism (edge water drive plus expansion of the secondary gas cap).

PVT and packed-column tests carried out at reservoir conditions evidenced the fact that miscible displacement with hydrocarbon gas is not feasible; on the contrary, first-contact miscibility with carbon dioxide exists.

Residual oil vaporization into dry natural gas, cycled throughout the reservoir, was experimentally studied and the results matched on a numerical model. A substantial increase in oil recovery by vaporization of the residual oil into cycled dry gas was evidenced.

A large number of surfactants and polymers were tested; a miscellar/ polymer process able to cope with the very high TDS content of the reservoir brine was designed and tested in the laboratory.

The drilling of a new well and the workover of four old wells resulted in a new phase of primary oil production from some pools of the field. The carrying out of an EOR pilot is pending on the termination of such primary oil production.

Title : Deep Water Trenching Vehicle TM-402	Project N° : 07.13/77
Contractor : TECNOMARE S.p.A. Address : S. Marco 2091 30124 Venezia - Italy Technical director (or person to contact for further information) :M.Rodighiero	Telephone N° : 708622 Telex : 41484

Tecnomare's TM-402 system has been designed and optimized for electrical cable, flexible pipe and flowline burying. It is directly derived from Tecnomare's TM-102 pipe burying system. The TM-402 system is composed of:

- An underwater, remotely controlled, tracked vehicle which supports the trenching tool and is capable of moving on the sea bottom guiding itself on the line to be buried.

- An umbilical cable for transmitting to the vehicle the energy necessary to locomotion, digging and other auxiliary functions and for receiving and transmitting the signals and the commands necessary for control.

- A surface set-up consisting of a motor generator for power supply to the system, a control room, an auxiliary power supply unit, and an umbilical cable handling module.

The basic characteristics of the vehicle are:

- Length in transport	5,6 m
- Length in operation	10,0 m
- Width	5,6 m
- Height	4,0 m
- Operating weight on land	22,0 t
- Operating weight on sea bottom	13,0 t
- Depth range	0:160 m
- Trenching speed (depending on soil condition)	10:400 m/h
- Trench width	0,25:0,40 m
- Trench depth	0:1,50 m
- Diameter of the line to be buried	50:300 mm

The vehicle is capable of operating on different soil types:

- Sand   : no limitation
- Clay   : no limitation
- Rock   : up to 150 kg/cm^2 compressive strength

Trials on land were carried out in Winter 1979 — Spring 1980.

Extensive and successfull sea trials have been carried out offshore Toulon (France) in the period June/July 1980. An electrical cable has been buried in a 22 m water depth. A new set of trials is planned before the end of 1981 in order to test the system for flexible pipe and flowline burying.

Title : Underwater Trenching Machine for Pipelines	Project N° : 07.14/77
Contractor : SAIPEM	Telephone N° : (02) 53531
Address : S. Donato Milanese 20100 Milano — Italy Technical director (or person to contact for further information) : M. Gioielli	Telex : 31246

    The research programme to develop the pipeline trenching machine will be carried out with the intention of studying and verifying the operation of individual components before proceeding with the development of the prototype.

    After having assembled in our logistical center at Cortemaggiore the necessary testing equipment, we have tested individual components such as evacuation pumps and samples of our cutting wheel with various cutters in different configurations. We have succeeded in verfying the validity and efficiency of various rotating systems and hydraulic bearings. Having reached these conclusions, we looked at the resources of various trust-worthy contractors and commissioned a feasibility study. Because of a change in market conditions this project was stopped in December 1980.

Title : Medium Range Circular Position- ing System	Project N° : 07.15/77
Contractor : SERCEL Address : Av. de Bel Air 44470 Nantes–Carquefou Technical director (or person to contact for further information) : M. Hythier	Telephone N° : (40)491181 Telex : 710695 CARQF.

Merops is a precise and non-ambiguous system of radio location which operates in the W.H.F. band and which has been developed to aid the positioning of ships, barges or platforms. It constitutes an extension of the SYLEDIS system which has been widely used in offshore operations since 1975.

Since SYLEDIS is limited in range to between 100 and 150 km. The techniques developed by MEROPS assumes a positioning precision of 10 – 20 m at a range between 250 – 400 km, night or day and in all types of weather. This accuracy is obtained despite the considerable adverse conditions which the wave propagation encounters at the horizon: large attenuation, fluctuations in level and propagation time, effects of abnormal refractions.

The MEROPS system embodies:

- A power amplifier (20 – 320 watts)

- A filter section which excludes on essential components of the spectrum (satisfying the most stringent national regulations: level of 5 microwatts out of band).

- A series of "anti-refraction" antennas equiped with a means of electronic commutation.

- A complementary means of statistical filtration of position data (U.C.M. and filter).

A permanent chain with large coverage is presently being developed in the Gulf of Mexico (from Louisiana to Mexico).

Title : Navigation System for Prospecting	Project N° : 07.16/77
Contractor : PRAKLA-SEISMOS GmbH Address : Postfach 4767 Haarstrasse 5 - D 3000 Hannover 1 Technical director (or person to contact for further information) :M.Schimech	Telephone N° : (0511)80721 Telex :

An integrated navigation system was developed for the prospection of hydro-carbon deposits off the continental-shelf regions, which uses doppler-sonar/gyro compass, NNSS and LORAN C as sensors. In the areas named above doppler-sonar works only in the mode water track. Because of the natural movement of the water (current) the movement of the ship in water track can be determined correctly, but not movements over long time against the sea bottom. The movements of the ship over the sea bottom are determined by including the velocity factor which can be derived from the LORAN C. The mean of these signals is stable but the signals are very instable in brief movements. The then remaining long-time errors are recognized and reduced by the NNSS system, whose data - converted in geographical position - improve the dead-reckoned positions. LORAN C was selected with a view to sea regions at the continental shelf of the European border seas, preferably the Norwegian Sea and the Greenland Sea. These regions have water depths up to 1,500 m.

Substantial performance features of the realized solution are:

- Independent navigation sub-systems, whose raw-data are combined in real-time to the most probable position over a process-computer.

- Simple system operation by interactive communication between operator and system.

- Programming predominantly in higher language and use of an effective real-time "core-only" operating system.

- Modular, noise unsensitive interface to sensors.

- Auto-test of the system by using the intelligence of the examinee.

The programme system was first tested in extensive simulations before performing the onboard tests.

The hardware of the system proved to be reliable in the rugged conditions of the North Sea. Higher surrounding temperatures and humidity which we encountered on unprotected board installations during our tests in the Mediterranean Sea have somewhat affected the system with regard to corrosion of metal parts, but have not affected its function.

Title : Deep Sea Connecting Techniques and Pipelines	Project N° : 09.07/77
Contractor : GERTH Address : 4, Av. de Bois Préau Rueil Malmaison — Paris Technical director (or person to contact for further information) : M. Leblond	Telephone N° : 749.02.14 Telex : 69066F

In each of the areas below the following work was carried out:

Laying:

— Optimisation of the RAT method.
  The programme of work has been completed. The optimisation study of the RAT method for industrial application that the following application, without being universal, are possible:

  a) the development of satellite deposits
  b) for the development of a geographical hydrocarbon sector separated from the main work–site

  and that its cost is competitive in certain cases only.

— "J" lay method.
  The following points have been studied in the framework of the flexibility study on J laying methods.

  a) general study of the method
  b) study of the dynamic behaviour of the pipe during laying
  c) study of connection method – choice of electron beam welding
  d) carry out metalurgical studies to show the suitability of various steels
  e) Pre–project machine
  f) component testing (joints, ramp).

Repair:

— Mechanical Connectors.
  The studies and tests carried out in this area can only be considered as an approach to mechanical connection. They have demonstrated the feasibility of a mechanical connection with air–tight metal to metal contact by means of a mechanical bearing on a formed collar.

The work envisaged has been completed. Complementary work is necessary to accomplish the applicability of this type of connection to pipes of various dimensions and steel types.

97

Preparation of the repair.

a) study and development of a system of acoustical methodology
b) development of tools to prepare the extremity of the pipeline (removal of concrete, removal of the coating, cutting).

Title : Waste Heat Recovery System Thermobloc	Project N° : 10.12/77
Contractor : BORSIG GmbM Address : Postfach D 1000 Berlin 27 Technical director (or person to contact for further information) : Lippert	Telephone N° :(030)4301-1 Telex : 18204

The project involves the realization of a waste heat recovery system.

The prototype used here is made up of conventional components (boiler, turbine, pump, piping systems) in such a way that mechanical energy can be recovered from the waste gas of a gas turbine via a close cyclic process (Rankine cycle).

The system extends and supplements an existing natural gas compressor station. The energy recovered in the system is used to drive an additional compressor which in turn assists the existing compressor or is used to make a further increase in capacity. The project is presently being realized in Canada, Province of Alberta.

Based on engineering performances and detailed planning in 1979, and after completion of corresponding contracts with the final customer and user of the proposed system after brief processing of all documentation, we were able to start and complete before the winter 80/81 the preparatory activities on site such as foundation laying, civil engineering, steel structures for platforms and pipe bridges, as well as the erection of additional buildings.

This warranted delivery and erection of all equipment in the spring of 1981 (planned schedule 1.4.1981) as well as the execution of all further erection activities during the summer of 1981. Parallel to this, orders for machines, apparatus and other items were placed. These items were delivered free on site before 1.4.81, which meant only one single assignment of heavy erection tackle was required. Erection and startup of the system is scheduled for the summer/autumn, 1981 so that the system can assist the compressor station during the winter period 1981/1982.

Title : Verolme LNG Carried	Project N° : 12.04/77
Contractor : VEROLME Address : Blaak 101 — Rotterdam Technical director (or person to contact for further information) :Dr.A.K.Winkler	Telephone N° :(010)112670 Telex : 26054

Part 1: Ship

Since due to its cryogenic containment system and the shallow draft required for the ports, canal of Suez, etc. the hull has a special geometry,
basic development work is required together with model testing. The development work for single screw ships has been made and confirmed by the model
tests, for a twin screw ship is till under way. The "Netherlands ship model
basin" in Wageningen conducted propulsion tests for a single screw carrier.
For this purpose a ship model was manufactured to a scale of 1:42 according
to the drawings (lines plan, body plan, etc.) made by us. The model was
modified later as a consequence of the initial testing. The results of these
model tests were extrapolated to full scale values giving the horsepower
required for such a carrier. Similar tests are being carried out for a twin
screw carrier. The tests are still under way and results therefore are not
yet available.

The second matter to be model tested is the manoeuvrability which is an
important element for the access to the European terminals. A first group
of such tests was made by Wageningen using their ship—handling simulator,
conslusion: "... no situation have been found, in which these ships could
not be controlled due to unacceptable manoeuvring characteristics...".

Part 2: Cryogenic system

For a very large LNG carrier (VLGC) of 330,000 m^3 LNG capacity the existing
cryogenic systems are not anymore suitable, moreover a main objective for
the design is the safety aspect. The cryogenic system under development is
a multi-container—system. A study was made by the "Netherlands bureau of
Industrial Safety TNO" concluding that "the risk of a "Verolnave"
330,000 m^3 LNG tanker would be 30 to 50 times less than of a "Conventional"
125,000 m^3 LNG tanker of the membrane type".

The development work included:

- design of the system including structural and other calculations
- material investigations taking into account cryogenic temperature
  conditions and especially fatigue resistance
- fabrication and assembly in the ship.

Design of the system

The classification and the admittance to a national port depends on the approval by a classification society and a national port authority. For this reason contact has been established with Lloyds register London, the American Bureau of Shipping as society and with the Netherlands Scheepvaartinspectie and the United States Cost Guard being port authorities. Lloyds is also providing engineering and use is made of their computer facilities, programmes and know-how. Two design alternatives were proposed of which Lloyds has accepted one solution and this solution is now being developed.

The results of the development of the cryogenic system are listed as follows:

- general arrangement
- cylindrical container layout
- container details
- supporting grid
- load and discharge diagram
- piping - general arrangement

Lloyds has accepted the layout of the cryogenic system.

Thermal study determining heat distribution, heat losses, boil-off etc. In the ship compartment where the cryogenic system is installed.

Report on constructional calculation of the heaviest loaded vessel.

Report strucan - pipe (baks report)

Design review (Lloyds report N° 80)

Completion of stage 1 and design review covering:

- design criteria of vessel
- containment vessels - struc. design
- tank support grid framework
- containment system
- insulation and thermal study results.

Computer calculation heaviest loaded vessel (baks report)

Double bottom grillage (Lloyd report)

Structural analysis of tank support structure (Lloyd report)

Stage 2: Model testing of the cryogenic system

Although involved structural computer calculations are being made, testing under static and dynamic conditions is required to confirm these calculations. The testing has been done in cooperation with Det Norske Veritas, Oslo, the Norwegian Ship Classification Society, who have the necessary model testing facilities and due to the development of the spherical tanks for LNG carriers the necessary testing know-how. The tests are being studied.

A preliminary study has been made to investigate sloshing and vibration of the cryogenic system with Det Norske Veritas. The result is satisfactory.

Title : Storage of LNG in Salt Caverns	Project N° : 14.02/77
Contractor : RUHRGAS/KBB	Telephone N° :(0201)1841
Address : Postfach 28 Huttopstrasse 60 - D4300 Essen 1 Technical director (or person to contact for further information) : F.Zündel	Telex : 0157818

In 1974 the possibility in principle of storing low temperature liquids has been experimentally proven in a test cavern of about 1 m³.

In the framework of the project the thermodynamic working properties of LNG in the cavern and the unstable 3—dimensional temperature field in the neighbourhood of the cavern has been established. Conformity with experimental results was satisfactory. Between calculated and observed rock displacement a qualitative conformity has been obtained.

The forming of cracks orthogonal to the cavern surface in salt rock has been clarified by extrapolating the known geophysical data of the salt. It has been shown that because of the insufficient knowledge of the material properties of the rock salt at low temperature and the complex mathematical boundary and starting conditions utilizing the available computational method no precise predictions regarding thermodynamic and rock mechanical behaviour in large caverns can be made without intensive experimental research.

The unstable temperature in the neighbourhood of the cavern for the storage of LNG has been calculated with a model cavern of 5,000 m³. The basis was a quasi—stationary storage at high pressure without discharge of boil off gas.

Some larger cracks in the cavern wall which developed during the tests did not influence the operational ability of the cavern. In practice it is important that crack forming should stop at a certain time and distance from the cavern wall.

Because of the high material streams involved in the rapid loading of LNG tankers it is necessary that LNG will be charged in the same time using a greater number of tubes. A storage system with 7 caverns has been developed, flow diagrams have been prepared and working procedures developed. A detailed economic investigation has been commenced to yield a qualitative indication of the comparison between salt cavern LNG storage and a conventional one with storage tanks. Salt caverns at the coast and distant from the coast have been taken into consideration.

Title : Development of a Total Struct- ural Monitoring System for Off- shore Platforms	Project N° : 15.03/77
Contractor : INSTITUTE FOR INDUSTRIAL RESEARCH AND STANDARDS, DUBLIN - IRELAND Address : Ballyman Rd. Dublin 9 Technical director (or person to contact for further information) : M. Benville	Telephone N° : 370101 Telex : 25449

As detailed in the contract work programme and the initial project proposal the project involves the installation of a wave and vibration data acquisition system on the gas production platform Alpha off the South Coast of Ireland. It is a steel template platform of conventional design in almost 100 m of water.

A contract was signed with Atkins Research and Development, U.K., to use their structure analysis package FATJACK. Work to date has been in defining the platform jacket in terms of nodal geometry and material properties. This data will be fed to Atkins computer to determine natural frequencies and mode shapes. The structure co-ordinates (200 nodes) and member characteristics have been loaded into the FRANLAB CDC computer, Paris via the Atkins installation in Epsom, U.K.

At present a wave data acquisition system is installed on platform Alpha off the South Coast of Ireland. This system comprises a Datel cassette recorder and IIRS controller, and is connected directly to a Marex wavestaff. Data from the system are processed on the IIRS computer in Dublin and the results are stored for long term analysis to produce statistics on such quantities as wave height exceedence and zero crossing periods.

One option available to us as regards vibration measurements was to record acceleration signals at pre-determined locations on the platform via accelerometers, the resulting outputs being recorded on a Nagra IV SJ tape recorder. Eventually it was decided to record both waves and platform accelerations on an integrated system with nine channels for data input — eight channels for the signals from eight accelerometers and one channel for wave data from the Marex wavestaff, with the option of increase in data channels if required for extra parameters such as wind and currents.

An order has recently been placed with a U.K. Company, Microconsultants Ltd., for supply of such a recording unit. It is a programmable computer, which as such will enhance our recording capabilities. It will be possible, on site, to programme the data acquisition system to record at required intervals and for required periods. Signals from the accelerometers and the wavestaff will be fed to the on-site computer via cables and recorded on magnetic tape for subsequent analysis in Dublin on the PDP11/34, to determine the environmental loading on the platform and resulting platform response. Actual natural frequencies, mode shapes and displacements calculated from collected data will be compared with corresponding results produced by the theoretical analysis. The theoretical model of the platform will subsequently be improved so that theoretical and actual dynamic properties coincide.

1978

4th Round Projects

Title : Correlator Stacker Demultiplexer CS 2502	Project N° : 01.11/78
Contractor : SERCEL Address : Rue de Bel Air — Carquefou F 44470 Nantes Technical director (or person to contact for further information) : M. Hythier	Telephone N° : 40491181 Telex : 710695 CARQF.

The project was dedicated to Research and Development of a Correlator Stacker for a large number of channels mainly designed to operate in VIBROSEIS TM geophysical crews.

TM Trade mark of CONTINENTAL OIL COMPANY.

The development of a prototype system was realized within a three year period 1978–1980. The first CS 2502 field qualification test was carried out early 1981 and commercialization of products started in July 1981.

CS 2502 is a peripheral equipment of a seismic data acquisition system to perform in real time, signal compression and vertical stack.

The system main characteristics are summarized below:

- high capacity: up to 500 channels at 4 milliseconds sample rate
- depth of investigation: up to 6 seconds at 2 or 4 ms
- full precision correlation: correlation and stack done in 32 bit floating point format
- unlimited acquisition length: real time correlation before stack does not fix any limit on vibration length
- noise elimination: editing of impulsive noise independant and self adaptive for each channel.

CS 2502 interfaces for SN 338 and SN 348 acquisition systems are available. Other systems interfaces are under design.

CS 2502 was designed around a specialized CSP processor able to handle 2 channels at 1 and 2 ms and 4 channels at 4 ms.

A CS 2502 is composed of:

- one central processor module with operator's console
- one to eight CSP modules according to the number of channels to process.

Title : Transverse Seismics	Project N° : 01.12/78
Contractor : GERTH	Telephone N° :749.02.14
Address : 4, Av. de Bois Préau 92502 Rueil Malmaison – Paris Technical director (or person to contact for further information) : M. Leblond	Telex : 69066F

Use of P wave seismic reflection has so far been the main exploration tool, leaving the possibilities of S waves unexploited.

S waves can provide additional information where P waves are absent.

It hence appears that combined S and P wave studies would enable new stratigraphical types of trap to be detected.

A seismic campaign was carried out on the SOUDRON field currently being exploited (Paris Basin). It consists in repeating with S waves a mesh already made with P waves and then calibrating on the basis of existing drillings.

The differences that appeared in the seismic calibration between S waves and P waves did not reveal the variations in the facies that were expected.

Lithological changes were revealed by this method for objectives situated at a distance of 1,400 m, but not for objectives at greater depths. Additional work will be needed to explain a certain level of scatter in the results.

Title : Development of a Seismic Technology for Improving the Resolution of Infra—Saliferous Petroleum Problems in the Mediterranean	Project N° : 01.13/78
Contractor : GERTH	Telephone N° : 749.02.14
Address : 4, Av. de Bois Préau 92502 Rueil Malmaison — Paris Technical director (or person to contact for further information) : M. Leblond	Telex : 69066F

With the development of the search for deep offshore petroleum deposits, the need to perfect methods for obtaining reflections from below s liferous intrusions (diapirs) became urgent.

An experimental campaign was conducted in the Summer of 1980 in deep Mediterranean waters (2,500 m) on a zone where the presence of diapirs had been recognized.

The campaign, which was initially planned in 1979 with a Vaporchoc source and three streamers was postponed till later owing to the development of a new source known as the "flexible gun".

1,200 km of profiles were made with the WLP (Wide Line Profiling) technique, enabling dips perpendicular to the seismic profiles to be investigated.

The data acquired during this campaign is new being processed using specific DP programmes.

Title : Horizontal Drilling Technology for Enhanced Recovery of Oil	Project N° : 02.10/78
Contractor : GERTH	Telephone N° :749.02.14
Address : 4, Av. de Bois Préau 92502 Rueil Malmaison — Paris Technical director (or person to contact for further information) :M. Leblond	Telex : 69066F

Drilling horizontal drains enables both the productivity of wells and the final recovery ratio of the oil in place to be increased. In addition, it enables enhanced recovery methods to be applied to the field under optimum conditions.

The aim of this project is to develop means and technologies for creating drains which follow the profile of the formations for the greatest possible distance.

Two techniques can be used for horizontal drain:

- flexo—drilling in which the drilling shaft is flexible and does not rotate, offering considerable capacity for transmitting remote measurements and remote control signals,
- conventional turbo—drilling or rotary drilling, where inclinations of 90° had never been reached and held.

The development of special flexible lines enabling horizontal drains to be drilled by flexo—drilling has led to a technical dead—end.

On the other hand, and for the first time in Europe, a 270 metres long horizontal drain was successfully drilled using conventional drilling rods on the LACQ field in June 1980. Three cores from 8 to 10 metres in length were taken from the horizontal drain. In addition, just as with conventional oil wells, this experimental well was completed with steel casings, also of the conventional type, and measuring instruments were run down inside the drilling string.

Drilling of new horizontal drains will stress bottom—hole measuring techniques (well logging) and the problems involved in producing from drains (completion).

Title : Oil Shows Analyser for Drilling Sites	Project N° : 02.11/78
Contractor : GERTH	Telephone N° : 749.02.14
Address : 4, Av. de Bois Préau 92502 Rueil Malmaison — Paris Technical director (or person to contact for further information) : M. Leblond	Telex : 69066F

The development of an oil shows analyser for drilling sites meets a twofold objective. First, it enables oil and gas shows to be detected during drilling, and second, it enables the nature of the zones crossed to be detected, whence the interest of the analyser in preventing blow-outs.

The design, construction and testing of a simplified unit on two drilling sites proved satisfactory.

An initial prototype of a commercial version has been built.

Title : Construction of a Source Rock Analyser Operating on Pyrolysis and Usable on the Drilling Site	Project N° : 02.12/78
Contractor : GERTH	Telephone N° : 749.02.14
Address : 4, Av. de Bois Préau 92502 Rueil Malmaison — Paris Technical director (or person to contact for further information) : M. Leblond	Telex : 69066F

The nature and state of evolution of source rocks encountered during drilling are all items of information that must quickly be acquired to decide whether to drill a core and even whether to continue or interrupt an exploratory drilling.

This information is at present obtained in the laboratory by means of methods that take a considerable time to apply, which often detracts from their value.

By contrast, the purpose of this contract is to provide the drilling site with a true geochemical automated "mini-laboratory" enabling the data needed to take the decision to be acquired in real time.

Three modules have to be integrated to make a source rock analyser that can be used on the drilling site: a source rock analyser proper, an organic sulphur analyser and an organic carbon analyser.

The manufacture of a prototype source rock analyser has resulted in satisfactory operating tests which enable consideration to be given to marketing the analyser; provided minor technological modifications can be made.

However, it did not prove possible to construct an integrated analyser owing to the difficulties encountered in developing the sulphur and organic carbon modules, calling for additional work not covered by the contract.

Title : Electro-optic Link for Off-Shore Application	Project N° : 0213/78
Contractor : SOURIAU AND CO.-FILECA Address : 9-13, Rue Gl.Gallieni — BP 410 92103 Boulogne Bill. — France Technical director (or person to contact for further information) : A. CHESNAIS	Telephone N° : (1)6099200 Telex : 250918

The aim of the project is to study the feasibility of an electro-optic link between installations at the surface (ship or platforms) and fixed or moving underwater equipment in deep waters (TV cameras for instance).

Electrical power: 15 KVA with low losses (10%)
Bandwidth   : 5 MHZ for remote control signals and data transmission (TV)
Depth   : at least 1,000 metres, investigations up to 10,000 metres

The main difficulties to be overcome are:

— Waterproofness (for insulators, fibers) and corrosion
— Pressures: fibers should absolutely not be strained
— Mechanical resistance (cable weight, low extension)

Numerous cable structures have been considered, consisting of from the center to the exterior:

— At the center: fibers laid out in steel pipes or in helical grooves along a rod
      Copper conductors 4mm^2 cross-section forming a septenary with the fiber pipes or in concentrical tapes

— A pressure structure made of hard steel wire layers, juxtaposed, around the central structure, or of several layers of steel tape winding

— A traction structure made of Kevlar or steel wires

— A plastic waterproof envelope

— and a wear sheath

Signal transmissions will be possible without repeater over a 10 km distance by means of optical coded impulses. The study of terminal connectors led to a solution where optical signals are converted into electrical signals in the plugs at cable ends, the junction with the sockets being electrical. The great difficulties met to ensure sealing at the ends of the cable lengths did not allow us to definitely settle for a cable structure, but the structure with fibers around a central rod and steel tapes winding to protect the fibers from the pressure, seems very promising and allows us to envisage a link without repeater to a depth which will be determined more precisely by new tests but which will certainly reach a few thousand metres.

The duration of the study will be extended by at least one year.

Title : Increased Production Capacity by TFL Techniques	Project N° : 03.57/78
Contractor : THE BRITISH PETROLEUM COMPANY LIMITED Address : Britannic House Moor Lane — London EC24 9BU Technical director (or person to contact for further information) :M. P. Smith	Telephone N° :(01)9208000 Telex : 888811

Following the successful completion of the Phase I Through Flow Line (TFL) evaluation work early in 1980 covered under EEC Contract N° 03.39/77 a new EEC Contract N° 03.57/78 was signed to support Phase II of the project.

Phase II of the project consists of the following:

1. The development and testing of a 4" x 2" carrier tool system to allow the use of larger diameter flowlines to satellite wells.

2. The development and testing of a multiway selector to divert tool strings to any one of several wells in a manifolded subsea completion.

Modifications to the BP Eakring test facility to incorporate the carrier tool system were completed in September 1980 and a planned five month evaluation of the system then commenced. Up to the end of December 1980 work was progressing satisfactorily towards achieving the objectives set out in (1) above.

The multiway selector was placed on order with Cameron Iron Works in January 1980 and was scheduled to be installed in the modified Eakring loop by the end of 1980. Due to unforeseen manufacturing problems however, delivery of the selector was delayed.

Title : Production Systems for Liquefied Natural Gas and Associated Gas in the North Sea	Project N° : 03.58/78
Contractor : SALZGITTER AG & PARTNERS	Telephone N° :05341-21-1
Address : Postfach 411129 D - 3320 Salzgitter 41 Technical director (or person to contact for further information) :M. Holekamp	Telex : 954481

As part of the project, detailed plans and designs are to be worked out for the construction and operation of LNG/LPG production systems located on mobile platforms in the North Sea, upto a water depth of 300 m. The design work is performed by a group of companies comprising Salzgitter AG, Howaldtswerke Deutsche Werft AG, LGA Gastechnik GmbH and Lurgi Kohle- und Mineralöltechnik GmbH.

The project is a direct continuation of Research Project 03.29/76, which involved the development of various concepts for the overall systems for offshore production of liquefied natural gas and methanol.

The designs that are now available as a result of the project work carried out cover a single-line process plant, a jack-up platform with substructure, a semi-direct transfer system, and platform-integrated product storage.

In addition, design work on a double-line process plant with related platform structure, on a tension-leg platform and on a concept for exteral storage was carried out.

In order to define the motion behaviour of an articulated tower that is to be used as supporting structure of transfer systems in deeper waters, extensive preliminary investigations were carried out for the purpose of model tests. These tests will be conducted in the spring of 1981.

The theoretical calculations that served as a basis for the design of the tension-leg platform were verified by means of a model test conducted by the "Schiffs- und Versuchsanstalt" in Hamburg. The obtained test results have confirmed the principle developed by the Project Group. In order to optimize the hydrodynamical behaviour of the system, additional tests using a modified subsea casing will take place early in 1981. In connection with the tension-leg platform, design work on the required riser connections from the sea-bed to the process platform has been started.

Apart from the completion of the detailed design for a double-line LNG/LPG plant, future project work will focus on finding a solution to the problems regarding product storage and transfer in the case of a tension-leg platform.

Title : Insert Wellhead Completion System	Project N° : 03.59/78
Contractor : SIPM B.V. Address : Postbus 162 Carl van Bijlandtlaan 30 Den Haag Technical director (or person to contact for further information) : O. Martinsen	Telephone N° : (70) 776655 Telex : 31005

This project is the first attempt to reduce the risk of damage to a subsea wellhead by reducing the height of the wellhead above seabed.

The project entails the installation of the wellhead, tubing hanger, master valve block and part of the flow loops inside the marine conductor of an underwater well. This effectively ensures that wellhead pressure integrity is maintained in the event of damage caused by trawl boards, nets or other objects.

The completion is a full three inch TFL (Through the Flow Line) completion with side pocket flow controls and accessories allowing pump down servicing of the well from a platform.

The design work on the wellhead and downhole completion components was finalised towards the end of 1978. After manufacture, acceptance testing together with a complete stack-up was carried out by the manufacturers in the U.S.A. The equipment was accepted by SIPM in March 1980.

Subsequently the equipment was shipped to Brunei and underwent a second stack-up and land test before installation offshore. This further test increased the confidence in equipment performance and introduced the equipment and running procedures to the local installation team.

Drilling of the well by the semi-submersible drilling rig SEDCO 135-E commenced mid-November 1980 and the well was ready for installing the Insert Wellhead Completion System by the end of 1980.

Title : **Design, Construction and Field-Testing of a Surface – Controlled Subsurface Safety-Valve System in Oil and Gas Wells**	Project N° : 03.60/78
Contractor : SIPM B.V. Address : Carel van Bijlandtlaan 30 Den Haag Technical director (or person to contact for further information) :O. Martinsen	Telephone N° :(70)776655 Telex : 31005

The project is aimed at developing an improved subsurface safety valve for protection against uncontrolled flow from producing oil and gas wells. Such a safety valve is particularly important for offshore fields where large numbers of usually highly productive wells are concentrated in single locations, and where uncontrolled flow can have serious consequences.

The improved safety valve is controlled from the surface and installed at the bottom of the well. Surface-controlled valves do exist but cannot be installed as close to the producing zone as desired because of limitations of their hydraulic control systems. The project therefore concentrates on the development of control systems for greater depth.

A study made during 1979 resulted in the selection of three control systems for joint development with the safety-valve industry.

I.   At the end of 1980 the design of a solenoid-actuated valve with an electric signal/control line to surface is near completion. Some key components have been tested and a field trial is planned.

II. Development work on a pilot-operated valve with an hydraulic line to surface has been initiated. Work is concentrating on the pilot valve. Tests with various prototypes are in progress.

III. Designs have been made of a valve controlled by pressure on the annulus between the concentric tubing and casing strings. As the subsequent designs were not considered optimal and showed no improvement, another manufacturer has been approached for this development.

Considerable attention is given to ensure that the systems are reliable and compatible with standard operations and equipment, since these are important parameters for economic application.

Title : Articulated Columns – Preparation for Industrial Implementation	Project N° : 03.61/78
Contractor : GERTH	Telephone N° : 749.02.14
Address : 4, Av. de Bois Préau 92502 Rueil Malmaison – Paris Technical director (or person to contact for further information) : M. Leblond	Telex : 69066F

The performance achieved when using articulated columns is an incitation to develop new technologies calling for experiments conducive to rapid industrialization.

In particular, the experiments involved the fatigue loads to which the components of articulated columns (riser) undergo as a result of their oscillations under the action of waves, winds and currents. The equipment was subjected to forces simulating five years of use in the North Sea. Tests were about to be completed at the end of 1980 and the results obtained have already shown that this equipment behaves satisfactorily with time.

Reduced-scale model simulation in the testing tank was performed so as to test the combination of a ship with a riser comprising several articulations, in deep waters. This experiment has led to the development that is now taking place of a design calculation programme for evaluating the loads generated by the combination of articulated riser and floating support and hence their certification by the requesite authorities.

Also, work was conducted on full-scale testing of rotary joints enabling several fluids to be transmitted at high pressure. An initial series of joints allowing service pressures of 60 bars was tested and work on testing of 200 bar joints has started.

Title : Deep—Sea Production Equipment	Project N° : 03.63/78
Contractor : GERTH	Telephone N° : 749.02.14
Address : 4, Av. de Bois Préau 92502 Rueil Malmaison — Paris Technical director (or person to contact for further information) : M. Leblond	Telex : 69066F

The medium term evolution of offshore operating techniques and examination of the numerous systems that can be envisaged for placing fields located in deep waters on production have led to study of the main factors common to the most realistic systems, based on maintaining the greatest possible number of drilling and production installations on the surface.

The work covered the following technological developments:

Riser in composite materials
Samples of composite material tubes were built with new glass fibres and carbon fibres. The initial results of the cyclic bending tests are now going on are encouraging.

Subsea manifold
Study of modular equipment for simplifying the maintenance operations required on the fluid distribution valves of manifolds has resulted in the construction of a prototype module.

Work inside the wells
Use of extensions in positioning pumped tools enables double completion inside wells to be avoided. The installation of safety valves at considerable depths (40 metres) below the wellhead for single completion has been successfully tested, using these extensions.

Safety — Reliability
Systematic analysis by the "faults flowchart" method employed in the nuclear industry has brought about improvement to the production riser, manifold and flowline connection systems.
Combustion of the associated gas during certain purging or emergency operations requires that a method of calculating the effects of thermal radiation be available. A mathematical model of this radiation where no wind is present has been made and tested. The effect of the wind will be studied in 1981.

Flowline connection and laying
A fully automated test for connecting and laying flowlines involving towing of the flowline a few metres above the sea bed has been studied so as subsequently to be able to perform sea trials in deep waters.

Title : Exploitation of Heavy Oil	Project N° : 03.67/78
Contractor : BRITISH GAS CORPORATION Address : 326 High Holborn London WC1V 7PT Technical director (or person to contact for further information) : K.W.S.Richards	Telephone N° :8316272 Telex : 261710

The Fluidised Bed Hydrogenator process is being developed for the gasification of all heavy oil fractions and is suitable for the processing of those heavy crude oils on or offshore which cannot be pumped without some treatment. The process is based on the direct thermal hydrogenation of the oil at elevated temperature and pressure.

Previous work had demonstrated that light crude oils could be gasified at the pilot plant scale. The present project is concerned with extending the range of feedstocks available to the process through further pilot scale trials and the identification of scale up parameters through the operation of a high pressure Fluidised Bed Development Vessel.

Seven pilot plant tests have been conducted to date using atmospheric residue feedstocks (SG up to 0.97) and an eighth using a bled of atmospheric and vacuum residue feedstocks. Operation for planned periods of up to 100 hours has been possible during which the oil feed rates have been increased to commercially acceptable levels. Some difficulties were experienced initially with the design of the oil injection system due to the nature of the feedstocks employed but these have been successfully overcome.

Supporting studies have revealed that between 75 and 85% of the sulphur in the residue feedstocks is converted to hydrogen sulphide. However, the high level of hydrogen sulphide within the hydrogenator causes severe corrosion of conventional alloys. Metallurgical coupons installed in the gasifier have shown that less conventional alloys perform better and these are being introduced as gasifier components where possible.

Construction of the Fluidised Bed Development Vessel has just been completed and commissioning is imminent. A range of specialised instrumentation techniques has been developed in order to measure the fluidisation quality of the bed and the mixing patterns of the particles.

Title : Design of a Heavy and Viscous Oil Production System (ROSPO MARE)	Project N° : 03.68/78
Contractor : GERTH	Telephone N° : 749.02.14
Address : 4, Av. de Bois Préau 92502 Rueil Malmaison – Paris Technical director (or person to contact for further information) : M. Leblond	Telex : 69066F

The ROSPO MARE offshore deposit is an accumulation of heavy oil amounting to about 200 million tons that can not be exploited economically using the conventional processes of natural drive or pumping.

Extraction of the oil from the bottom of the well to the point of removal is the subject of the present contract.

Although encouraging, the initial results reveal the difficulties involved in placing such a heavy crude on production.

As a result of short duration production tests made on ROSPO MARE 2 well, the following production system is envisaged:

- use of long-stroke pumping units and need to develop them for deviated wells,

- injection of surfactants at the separator to break up the highly viscous emulsion formed by water in the ROSPO MARE crude,

- transportation to the point of removal by accomplishing inverse emulsion, i.e. emulsion of ROSPO MARE crude in water, which offers the particularity of being less viscous than the crude itself,

- separation of the constituents of this emulsion at the point of removal.

The working programme marked time owing to the lack of data that long duration production tests would have provided for evaluating the possibilities of primary exploitation of this type of accumulation.

Title : Development of Columns not Sensitive to Motions	Project N° : 03.70/78
Contractor : LINDE AG	Telephone N° :089.72731
Address : D-8023 Höllriegelskreuth	Telex : 521 2725
Technical director (or person to contact for further information) : M. W. Förg	

The exploration and liquefaction of natural gases in deep sea waters on floating carriers requires research for the development of columns not sensitive to motions. Results obtained within different research programmes permit only a statement of the tendency of the behaviour of columns erected on offshore platforms.

To obtain a reasonable basis for the design of "offshore columns" it is necessary to test a distillation column (distillation being a most sensitive process) under motions simulating the motions of a floating structure in different sea states. Such stochastic movements (6 degrees of freedom) can be performed by a "motion simulator" at Det Norske Veritas, Oslo.

The results of the tests carried out on that simulator should enable:

- to determine the decrease of efficiency of columns under offshore conditions
- to determine guarantee figures for a process plant
- to determine the down-time per year of an offshore process plant caused by malfunctioning columns.

Up to the end of December 1980 a pilot plant for a distillation process was designed, manufactured and mounted at Linde, Munich. That means construction of a special plate column and a packed column (diameter 700 mm, model scale 1:4) and of a packaged unit (10 x 4 m) containing reboilers, vessels, pumps, piping, control room etc. All equipment was prepared for an easy transport and a quick assembly after arrival at Det Norske Veritas, Oslo.

Parallel to this work the performance of the tests on the motion simulator was prepared especially with respect to safety for operating an explosive and flammable test mixture used for the process and a moved column which is connected with the rest of the plant by flexible hoses (diameter 50 - 200 mm).

After testing the functioning of the pilot plant in Munich, all equipment will be shipped to Oslo in August 1981 and in September 1981 the test runs on the motion simulator will be started.

Title : Drainage System ROSPO MARE	Project N° : 05.09/78
Contractor : ELF ITALIANA	Telephone N° : 5896441
Address : Via I. Nievo 35 00153 Roma	Telex : 61483
Technical director (or person to contact for further information) :M. Etzbach	

Following the decision taken during the first half of 1980 to carry out a long term production test on the ROSPO MARE field, and without waiting for the results of this test, studies and research on drainage methods were partially interrupted.

The only current activity during the second half of 1980 consisted in resuming some of the 1979 mathematical simulations by means of a new implicit resolution model in order to explore the post-breakthough production period, which was not possible with the former model owing to numerical instability.

The hypothetical data used so far, the new model has demonstrated:

1. that for a vertical well with a flow of 250 m^3/d, the oil recovery following water breakthrough represents 60% of that prior to break-through.

2. that these results remain unchanged when the flow of the vertical well drops to 50 m^3/d, (the flow has no effect on the evolution of the WOR, at least as long as one remains in hypercritical range).

3. that for a horizontal well 500 m long with a flow of 250 m^3/d, the recovery after water breakthrough is only 13% of that before break-through.

4. that the relative gain in production of the horizontal well compared to the vertical well is 5 to 1 at water breakthrough, is only 3.6 to 1 for a WOR of 3.

These results therefore tend to reduce the interest in horizontal wells, though without eliminating them thereby.

In view of the low hope of success that we ascribe to the other methods envisaged (CO_2 injection, steam injection, or injection of gelling solutions) and the remarkable results recently obtained in horizontal drilling techno-logy (see the LACQ 90 experiment), we consider that the horizontal well method offers the most attractive line of research for combating water ingress and enhancing recovery on the ROSPO MARE field.

To conclude this study, we recommend that a horizontal well be attempted from the production test platform to be set up on the ROSPO MARE structure in 1981.

Title : New Techniques to be Applied in Piropo Oil Field to Achieve Economical Feasibility	Project N° : 05.10/78
Contractor : AGIP S.p.A. Address : 20097 San Donato Milanese 20100 Milano Technical director (or person to contact for further information) : M. Ciarichi	Telephone N° : 53531 Telex : 31246

A well-drilled back in 1975 in the Piropo structure, Adriatic Sea, evidenced the presence of a huge deposit of heavy (18°API) oil in fractured carbonates, Scaglia fm. No commercial oil production was obtained.

The purpose of this project is to evaluate different well stimulation and EOR techniques which show a potential capability of bringing the oil production rate to a commercial value.

In the first phase of the project the existing seismic data were reprocessed by the SISMOLOG method, to locate the most fractured area(s) of the reservoir rock, where a better productivity should be expected. The correct structural picture of the carbonatic unit, the porosity sequence, the thickness of the various layers and type of fluid present in each layer were evidenced by SISMOLOG reprocessing of the seismic traces.

Phase two consisted of drilling, coring, logging and testing a well (Piropo 2) located in the most promising area of the deposit.

The lithological sequence encountered by the well substantially confirmed the results of the SISMOLOG study.

Three DST and six production tests were carried out to locate the water-oil contact and to evaluate the productivity of the oil-bearing intervals.

The most promising oil-bearing zone was stimulated twice by acid jobs, without obtaining a sizable oil production. A massive hydraulic fracturing (MHF) job, with the injection of some 180 thousand gallons of fluids in 11 stages was performed, and a stabilized oil production rate of 11 cu m/day (70 bbl/d) only was obtained.

The study of the cores and fluids recovered from the well is underway, with the aim of finding a process capable to reduce reservoir oil viscosity, and thus substantially improve oil production rate from this well.

Title : Enhanced Recovery from Very Heavy Oil Reservoirs	Project N° : 05.11/78
Contractor : GERTH	Telephone N° : 749.02.14
Address : 4, Av. de Bois Préau 92502 Rueil Malmaison – Paris Technical director (or person to contact for further information) : M. Leblond	Telex : 69066F

Major heavy oil resources remain unexploited owing to the inherent limitations to thermal recovery methods that are usually applied to them.

This contract has as objective the development of techniques specific to reservoirs containing highly viscous oils or bitumens. On the industrial scale, it is intended to evaluate the potential of recovery methods combining injection of organic solvents and conventional thermal methods, and injection of hot fluids, with in-situ combustion.

The work concerned a range of 7 heavy crudes from a variety of origins and with densities of from 0.96 to 1.01.

The selection of the various organic solvents was made in vivo in the light of their influence on the precipitation of the alsphaltenes present in the heavy crudes and the drop in viscosity related at one on the same time to the dilution of the crude by the solvent and the action of the temperature.

From the standpoint of the in vivo measurements, a number of tests combining injection of solvent and in-situ combustion or injection of steam displayed significant effects.

Title : Electrical Process Applied to Enhanced Oil Recovery	Project N° : 05.12/78
Contractor : SYMINEX	Telephone N° : (91)529011
Address : 15, Bd. Cieussa F 13262 Marseille Cedex 2 Technical director (or person to contact for further information) : M.Kermabon	Telex : 400563

The research done by SYMINEX has shown that the injection of pulsed electricity in reservoir samples increases the recovery of oil.

The movability of oil appears to be enhanced by the process of electro-osmosis at the capillary level. The oil is moved from areas which are not normally flooded, towards channels with a better permeability.

A survey was made by the Ecole Centrale de Paris, to verify that the process is genuine and not the result of secondary effects which might interfere at the laboratory level (electrode polarisation, action of alkaline ions...)

Tests on rock samples helped to define the values of the electric field and energy levels at which the process is economical. Using these values to study electric field distribution in a reservoir, we find that the area submitted to effective electric treatment is as large as the area which is produced around the well. To generate this effective electric field, the power to be delivered at the reservoir level is 1 Megawatt. A feasibility study has demonstrated that delivering an electrical power of 1 MW at the reservoir level is technically possible (with a cable or with insulated tubing as a conductor).

For a field application of the process, a choice has to be made:

- should the current application and the production of the well be consecutive or simultaneous (the latter necessitating a more sophisticated system).

It looks as if alternating the two phases is feasible at the reservoir level. In the electrical chain, the critical point appears to be in the proximity of the electrode. The heating of this formation will have to be closely monitored in order to prevent electrolyte vaporisation.

It should be noted that the process can be applied to a great variety of reservoirs, but seems to be better suited to fractured reservoirs.

Title : New Offshore Hydraulic Pile Hammer	Project N° : 06.07/78
Contractor : BSP INTERNATIONAL FOUNDATION LTD.	Telephone N° :0473 830431
Address : Claydon Ipswich — Suffolk	Telex : 98115
Technical director (or person to contact for further information) :M.R. Storey	

B.S.P., recognising the problems of extracting oil and gas under conditions of increasing water depth, have developed a range of hydraulic hammers which are capable of operating underwater.

The development was centred on hydraulic actuators of 10 t and 20 t capacity. The simple design of the hammer chassis and the modular construction of the actuators allows any number of them to be used according to the size of hammer required. The actuators are easily replaced in the event of a failure so that spare units obviate the need for a complete standby hammer.

The hammer does not operate in an air bell, so small air hoses to the surface are unnecessary and the impact cap, which provides for the most efficient transmission of impact force through to the pile, does not need a separate hydraulic supply as is the case with nitrogen buffer cushions.

A pair of 10 t actuators mounted on a hammer have operated satisfactorily in 120 metres water depth and an electrical control system has been developed and proven at the same depth.

Other developments include a slender hammer which is able to pass through pile guides, removing the need for pile followers as is the case with above water driving.

Title : DAVID — A Tethered Submersible for Use as a Tool by a Diver	Project N° : 07.22/78 07.33/78
Contractor : ZF—HERION— SYSTEMTECHNIK GmbH Address : Postfach 1560 Stuttgarter Strasse 120 D—7012 Fellbach Technical director (or person to contact for further information) : M.K.Wiemar	Telephone N° :(0711)5071 Telex : 7254507

In initial estimates, it was anticipated that the project would be completed by December 1981, and this is now expected to be achieved by December 1982. The complete system "DAVID" is considered as two separate projects.

1. TH 07.22
   "Entwicklung von Steuerungen, Antrieben und Werkzeugen für den Unterwasser—Einsatz". (DAVID II)
   Development of controls, drive systems and tools for sub—sea operation.

2. TH 07.33
   "Entwicklung eines Taucherhilfsfahrzeuges für UW—Inspektions und Wartungsarbeiten (DAVID III)

   Project Status end 1980: In final stage of design.

Project Summary:

The project is concerned with the specification, development, manufacture and testing of the system prototype, and it is intended that the system should become fully operational in the North Sea by July 1982.

The system consists of a submersible vehicle weighing approximately 5,000 kg, a handling system for launching and recovering the vehicle, a control container with facilities for controlling and monitoring the vehicle activity and a diesel generator power supply unit. The system is designed for operation either from a platform or from a diver support vessel.

The submersible vehicle provides the diver with tools, power and facilities, which enable him to work with high efficiency and safety when carrying out tasks associated with the inspection, maintenance and repair of offshore installations. The vehicle can be guided and positioned under either local control by the diver or remote control from the surface. An integrated clamping system enables the vehicle to be attached to a structure and positioned at any desired angle. A moveable platform can be erected to provide the diver with a stable base from which he can work. Other on-board facilities include a range of hydraulic power tools, a power winch, equipment for NDT, water jetting and pumping equipment.

The initial detailed specification was prepared jointly by ZFHS, Ocean Consult and the Norwegian diving company Scandive in the autumn of 1979. This specification has been realized in the final design.

Prototype manufacture is expected to commence in August 1981 and initial testing is to begin early in 1982.

Title : Improvement of Subsea Exploration Technology in Great Depths	Project N° : 07.29/78
Contractor : GERTH	Telephone N° : 749.02.14
Address : 4, Av. de Bois Préau	Telex : 69066F
Technical director (or person to contact for further information) :M. Leblond	

Reconnaissance of the geological characteristics and mechanical properties of sea beds by sampling cores is of interest to all sectors of the oil industry involved in harnessing deep offshore oil.

The objectives of this contract are first to develop a sediment core driller for decametric penetrations and second to add a rock core driller to the CYANA bathyscaph, which is the only manned European device capable of operating in very great depths. At the same time, the need has appeared to improve the navigating rotation and data collection capabilities of the CYANA.

Core driller

A compact sediment core driller has been built and successfully subjected to sea trials. The results have enabled the construction of an industrial size (20 metres long) prototype to be started.

Cyana

The Cyana bathyscaph was equipped with a prototype rock core driller to extract cores 100 mm in length and 25 mm in diameter. Its implementation using a remote-handling arm was successfully tested at sea. A series of additional tests is also planned.

Substantial improvements have been made to the CYANA with respect to its navigating location ability, the identification measurements now being made from the vessel itself, and not from the support ship. The capacity for acquiring data on the environment has also been increased without any corresponding increase in the volume and weight of the equipment.

Title : Cryogenic Flexible Pipes	Project N° : 10.14/78
Contractor : COFLEXIP S.A. Address : 23, Av. de Neuilly F 75116 Paris Technical director (or person to contact for further information) : M.R. Renard	Telephone N° : 747.0530 Telex : 610303F

The main applications of cryogenic flexible pipe lines as described in this project started in 1979, are for use in L.N.G. transfer between liquefaction units, storage units, loading units, transportation, unloading and re-gasification.

Preliminary tests and studies were aimed at selecting the materials the most suitable for service requirements. The work has been essentially devoted to research on metal components and to the thermal insulation material.

The internal corrugated tube made of 304-L austenitic steel, is impervious to the cryogenic products while providing flexibility, resistance to fatigue (several millions of cycles) and resistance to external and internal pressures (a sample of 4" i.d. corrugated tube under a pressure of 2,000 psi has shown no visible damage).

Welding tests of the corrugated steel tube have been run in conjunction with some specialised companies. These tests have enabled the development of a welding robot suited to the adopted manufacturing procedures of the tube.

Comparison tests run on various foams have enabled the selection of a thermal insulation foam meeting with the required characteristics. Further tests conducted on this selected foam (dynamic, static tests, etc.) have proved satisfactory. The insulation foam layer will be protected against possible cracks.

Furthermore, computer calculations have verified the test carried out on the materials and structures.

Qualification tests will be run at our plant on samples of Coflexip 8", 12", and 20" i.d. cryogenic flexible pipes.

Title : High Pressure Flexible Piping Development	Project N° : 10.16/78
Contractor : DUNLOP LTD — OIL & MARINE DIVISION	Telephone N° :(0472)59281
Address : Moody Lane — Grimsby DN 312SP	Telex : 52184
Technical director (or person to contact for further information) : M.B.Eastwell	

The objective of this project is to develop high pressure flexible piping — predominantly for subsea flowline applications but bearing in mind, also, the coming need of flexible riser connections.

Existing technology satisfactorily meets the specification parameters of existing "low" temperature sweet wells but the requirements of E.E.C. governments to operate a responsible reserve depletion policy will demand a significant advance in technology to overcome increasing hostile physical elements being experienced at the wellhead.

This development programme is proceeding to develop high pressure flexible piping predominantly for flowline applications which can withstand:

1. Produced reservoir fluids/gases at elevated temperatures.

2. As above together with significant levels of gaseous H_2S.

3. Well stimulation/injection fluids or gases.

4. High fatigue conditions.

The programme is structured to solve three major problem areas, namely:

1. The stress factors.

2. Problems due to permeation and exagerated by high temperature.

3. Long term problems due to corrosion and exagerated by high temperatures.

By end 1980, tentative designs of pipe constructions had been developed to withstand pressures up to 700 bars working. (This requirement is necessary when T.F.L. tooling is employed). Minimum bursting pressures in excess of 1,750 bars are required.

A machine has been designed to manufacture these products, bearing in mind the necessity to produce these pipes with the utmost degree of quality assurance. (This was expected to be operational for prototype work by mid 1981).

The problem of permeability is central to the carrying of "live" crudes as well as stimulation gases. To this end, cylindrical "bombs" have been designed whereby composite structural samples can be subjected to gases or "live" fluids at temperature and pressure. They have the facility of being capable of taking H_2S. A programme of testing the range of polymers and plastomers was initiated during the last quarter of 1980 and is scheduled to continue throughout 1981 and into 1982.

Corrosion problems due to permeations of agressive materials or arising due to external damage of the product are hazards that must be guarded against. Laboratory rigs have been designed and constructed to test the corrosion resistance of all materials that may be used in these products and to evaluate the various ways in which protection can be added, this work — because of the need for "real time" testing will continue throughout 1981 and beyond.

Title : Offshore Loading System for LNG/LPG and Other Refrigerated Fluid	Project N° : 10.17/78
Contractor : SALZGITTER AG & PARTNERS	Telephone N° : (05341)21-1
Address : Postfach 41 11 29 D - 3320 Salzgitter 41 Technical director (or person to contact for further information) M. Haferkamp	Telex : 954 481

The working group Salzgitter AG, FMC Europe SA and Peiner Maschinen- und Schraubenwerke AG have designed and developed a transfer system based upon the use of the cryogenic Chiksan swivel joints combined with straight pipe lengths within a double diamond shaped articulated loading arm. The main features of the overall system are as follows:

- all metal arm composed of rigid pipe sections and swivel joints,
- the use of service proven cryogenic swivel joints that include double product seals with means to monitor seal performance,
- the use of a computerized sensing and control system allowing continuous monitoring and display of the relative position of the LNG carrier being loaded,
- a dry break disconnect system in case of emergency,
- a tensioning guidance system for ease of connecting/disconnecting,
- simplified on-site maintenance.

For this system, a model to scale 1:10 has been fabricated and provided with peripheral monitoring and control installations. Using a motion simulator, the model was subjected to a series of tests at Det Norske Veritas. The essential aims of the tests were as follows:

- for the connection phase, simulate the possibilities and limits of the human intervention,
- for the disconnection phase during which the human control will be restricted, check if the overall system behaves as it was foreseeen,
- for the loading phase, check the efficiency and reliability of the monitoring and control system.

The tests have been successfully completed, and a preliminary evaluation of the extensive test results has fully confirmed the static and dynamic stress calculations and computer simulation previously effected. The complete evaluation of the obtained data is due to be accomplished early in 1981. The classification of the overall system will probably take place by the end of 1981.

Title : Insulation and Barrier System	Project N° : 12.05/78
Contractor : SHELL RESEARCH U.K. Address : Shell Centre London SE1 7NA Technical director (or person to contact for further information) : M.H.L.Beckers	Telephone N° : 934-1234 Telex : 919651

Polyurethane Foam Development

The properties of the polyurethane foam have been established for the conditions under which it is sprayed. In addition the effect of water absorption into the foam and the effect of water vapour permeation into the foam were investigated. Tests on the effect of ageing on mechanical adhesion and on the fracture properties of the foam were also completed.

The polyurethane foam is being checked for its compatibility with the products of the inert gas used in gas-freeing ships' cargo tanks.

Laminate Development

There are two types of laminate in the system (see July 1980 Progress Report):

a) Laminates used to reinforce the polyurethane foam;
b) Channelled laminates which in addition to reinforcing the foam permit the circulation of inert gas through the system to detect possible leaks.

The mechanical and thermal properties of the reinforcing laminate have been established by tests. The testing of the channelled laminate under fatigue loads is now in progress. The channelled laminate located at the steel inner hull is intended for detecting leaks in the secondary barrier and also to prevent possible small cracks which might develop in the steel from propagating into the system.

This feature is being tested by investigating the adhesion of the channelled laminate to steel and the effect of water (from ship's ballast tanks) on this adhesion. The ability of the channelled laminate to bridge cracks is under investigation.

Manufacturing processes for the channelled laminate are now being finalised.

Quality Control

The application of the system (both polyurethane foam and laminates) requires strict quality control. Specifications have been written defining the necessary procedures and quality control tests to ensure that adequate and consistent material properties are obtained during application.

20 m^3 Test Tank

Testing of the system in the 20 m^3 tank has confirmed the viability of the design and of the polyurethane foam and laminate materials.

During the 20 year life of an LNG ship there will be 400 laden voyages during which the insulation system will be subjected to thermal and pressure cycling loads from the LNG cargo and pressure cycling loads from the ballast water, which acts upon the inner hull steel to which the system is attached. The system has now been subjected to the complete range of loads, simulating a 20 year ship life, with no signs of any damage or deterioration.

The tank will now be used to determine the effect of inflicted damage on the performance of the system.

Pump Support Rig (formerly Penetration Rig)

A half size structure of the pump support rig was constructed and tested under fatigue loading. The structure however failed at half the 20 year required life. An analysis of the failure is under way and the pump support rig will be redesigned on the basis of this experience.

500 m^3 Proving Tank

As mentioned in the previous Progress Report the installation of the system in this tank awaits:

a) Approval of the system by the Classification Societies.
b) Involvement of a third party to promote the system.

When these conditions are agreed the system will be installed in the tank under realistic commercial conditions.

Application Equipment

A structure on to which foam and application equipment can be mounted was completed in 1980. The foam and laminate application equipment which will be used in the 500 m^3 proving tank are now being finalised.

Title : Underwater High Density Energy Source	Project N° : 13.05/78
Contractor : COMEX, S.A. (FRANCE)	Telephone N° :(91)410170
Address :B.P.49 287 Ch. de la Maldrague Ville F-13314 Marseille Cedex 3 Technical director (or person to contact for further information) :	Telex :410985

Objective

Design, fabrication and test of an autonomous underwater power source, providing an higher energy density ratio than lead acid batteries: 35 Wh/Kg versus 19 Wh/Kg.

Power Source

STIRLING engine model P-40 providing 25 KW at 3,000 rpm associated to a 220/380 V 50 HZ AC generator.

Fuel

– Gas-oil and/or other hydrocarbons or alcohol
– Pressurized oxygen.

Energy supplied

– Electrical with 25 KVA/50 HZ AC generator
– Thermal: not less than 40 KW at full power
– Capability of hydraulic (direct coupling).

Underwater housings

Several containers are necessary:
– Engine pod including AC generator
– Associated battery pod
– Heat exchangers
– Electronic control system container
– Storage tanks for gas-oil and oxygen.

Applications

– Underwater power source for habitats, power packs, tools, motor pumps, etc..
– Power sources for diver lock-out or observation submersibles
– Tethered or untethered remote controlled vehicles.

Project Status

- Design and fabrication of engine pod
- Pod assembly in progress
- Test of basic components: combustion chamber and 30 bars exhaust system
- Complete tests planned for summer 81 (in air) and autumn 81 (in water).

Title : Hydrocarbon Storage	Project N° : 14.03/78
Contractor : SIR ROBERT MC ALPINE	Telephone N° : 01 837 337
Address : P.O.Box 94 Bernard Street — London WC1N 1LG Technical director (or person to contact for further information) : R. Clare	Telex : 22308

A feasibility study has been carried out on a single containment concrete hydrocarbon storage system.

The following items have been examined:

1) Safety aspects and environmental impact of cryogenic storage.

2) Survey of past and existing cryogenic tanks.

3) Construction materials for cryogenic storage.

4) Design features of a cryogenic storage tank.

5) Storage at ambient temperature and pressure.

6) Siting of storage facilities.

7) Market potential for concrete tanks.

8) Preferred design for cryogenic storage.

The work carried out thusfar has demonstrated the feasibility of a single containment hydrocarbon storage system of concrete construction. Phase 2 of the programme will develop a detailed design of such a system and will also develop some first order costs.

Title : Self Inst lling Subsea Storage System	Project N° : 14.04/78
Contractor : E.M.H. — HALCROW — EWBANK	Telephone N° : 602 1122
Address : 29, rue de l'Abreuvoir F — 92100 Boulogne Technical director (or person to contact for further information) : J.Alleune	Telex : 204 586

The project develops an original design for storing and loading oil on tankers in offshore oil fields. The design in a combination of the proven and successful articulated loading column with a prestressed concrete subsea storage tank. Project studies the construction, transport, installation and exploitation of the system.

Development in a representative design condition is carried up to a point where the design is suitable for commercial application.

Progress has been made following three general directions solving the most unclassical problems raised by the project.

- Installation procedure: one procedure has been selected among three. Theoretical work and computer programme to simulate the static installation of the system have been made. Final ballast arrangement in the tank and model test specifications at scale 1/88 are being defined.

- Tank design:one optimum tank section has been choosen. Theoretical work and computer models have been made to assess the design loads on the tank during tow—out, installation and exploitation (foundation design, creep—temperature effects).

- Oil — water diagram: one oil — water separation system has been selected to suit the system characteristics. Flow diagram for oil — water pipings in the tank and column have been finalized and their arrangement are being defined.

At this stage, all technical tools, design procedures and system features have been choosen. Final ballast arrangement in the tank is being sought to fulfill structural and installation requirements, prior to starting model testing.

143

Title : Development of Unlined Concrete Storage Facilities for Liquefied Natural Gas	Project N° : 14.06/78
Contractor : **TAYLOR WOODROW CONSTRUCTION LTD.** Address : 345 Ruislip Road Southrall—Middlesex UB1 22X Technical director (or person to contact for further information) : R. Browne	Telephone N° : (01)5782366 Telex : 24428

The programme is aimed at fully developing the potential of concrete as a material for either primary or secondary containment of LNG. This has required an extension of available expertise in both concrete technology and prestressed concrete design from elevated temperatures, which was applicable to nuclear structures, down to temperatures as low as −165° for LNG storage.

The initial phase of the programme involved a comprehensive survey and analysis of over 100 references related to the behaviour of concrete and steel from 0 to −196°C and the design of LNG tank. Following this analysis, 20 different concretes were tested to select those which achieved the characteristics required for cryogenic storage, i.e. low permeability, high ductility for reduced crack proneness, and resistance to thermal cycling.

Four concretes, selected from the original 20, are to be more comprehensively tested. This will provide data on the engineering properties of concrete down to −165°C, required for the design of LNG tanks. Reinforcing steel and prestressing systems are also being evaluated.

The final phase of the work will involve the construction, commissioning and testing of a model storage unit. It is intended that this will demonstrate that site constructed structures can achieve the same basic properties which have been found in the laboratory in the earlier stages of the programme.

Title : Diagnostic Methods for Offshore Structures	Project N° : 15.06/78
Contractor : **TECNOMARE** S.p.A. Address : S. Marco 2091 30124 Venezia Technical director (or person to contact for further information) :M. Rodighiero	Telephone N° : 708622 Telex : 41484

Purpose of this research project is to develop the technology (hard-ware and software) necessary to know the status of health of offshore structures by means both of occasion 1 integrity checks and of a continuous monitoring. Furthermore the project investigates the implication of a "damage-tolerant" approach in the design-monitoring of offshore structures. This approach, mainly based on fracture mechanics, allows to predict crack propagation both in design phase and during service life; in this last case the actual stress history obtained through a net of strain gauges installed on the structures is used.

The main features of the project are the following:

- Analysis and selection of the most promising detection methods.
- Model lab. tests for examining in detail the real potential of vibration monitoring and acoustic emission as detection methods.
- Set up of computer procedure for the optimization of a measurements campaign with the Vibration Monitoring instrumentation and for the analysis of the experimental data.
- Study , construction and test of a portable instrumentation system, for occasional structural checks, based on V.M. techniques.
- Study and basic design of a permanent instrumentation system based on a combination of different subsystems.
- Development of a computer programme based on fracture mechanics, capable of predicting the crack growth as function of the stress hystories obtained by a net of strain gauges or theoretically predicted in the design phase. It has to be pointed out that such an approach allows a reduction in inspection, as it will be required only when the maximum allowable damage is theoretically accounted.

At the present, the analysis of the most promising methods, the V.M. and acoustic emission laboratory tests and the set up of V.M. optimization and damage localization computer procedures are completed. The Vibration Monitoring portable instrumentation system is under construction. The basic design of the complete permanent instrumentation system is under development

Title : Meteoceanographycal and Structural Data Acquisition System to Improve Offshore Platform Design	Project N° : 15.07/78
Contractor : AGIP S.p.A.	Telephone N° : 53531
Address : 20097 San Donato Milanese 20100 Milano	Telex : 31246
Technical director (or person to contact for further information) : M. Magni	

During the last few years the exploitation of offshore reservoirs has evidenced the need of increasing the platforms dependability either mobile or fixed, by refining the design methodologies, which become more important due to the continuous dimensional increase of this type of structures.

The data acquisition system installed on "BARBARA" platform located in the Adriatic sea some 57 km N.E. Ancona, has been conceived in order to achieve the following objectives:

- To verify the scattering of the results of present calculations either in the static or dynamic mode, from actual measured values. This confrontation should allow a more realistic evaluation of the actions transmitted by the sea to the submerged structures.

- To calculate the fatigue damage of the structural members, on the basis of the load cycles actually supported up to a certain moment, to evaluate the suffered damage at that moment and to make previsions on the remaining fatigue life.

- To individuate irreversible failures of the structure (welding cracks, foundation settlement) and to plan, in an optimized way, the periodical visual inspections (by means of a spectral analysis of whole records obtained at the moment).

The system is programmed to carry on the following operations:

- Environmental condition measurements, measurements of the structural stresses and measurements of the dynamic response of the platform.

- Transmission, recording and pre-elaboration of the collected data.

To date (31.12.80) the whole equipment has been installed and it is being checked on the jacket, while the deck is still under construction. It will be installed on July, 1981, at the end of drilling operations being now performed.

Title : Development of Measurement System to Optimize Computer Programmes and to Monitor Dynamic Behaviour of Offshore Platforms	Project N° : 15.08/78
Contractor : SYMINEX	Telephone N° : (91)529011
Address : 15, Bd. Cieussa F-13262 Marseille Cedex 2 Technical director (or person to contact for further information) : .A.J.Kermabon	Telex : 400563

The project began in 1978, and deals mainly with the measurement and monitoring of the dynamic behaviour of offshore platforms. The final purpose of this research is to develop methods and corresponding equipment to monitor the structural integrity of platforms by forced vibrations analysis.

On the 1st January 1980, theoretical and experimental research results allowed us to define two methods, one for the detection and localization of cracks, and the other to evaluate the loss of integrity of the whole platform resulting from the damage. The equipment for the application of these methods offshore has been designed and realized.

During 1980, this equipment was tuned and then used to test the two methods offshore. A bracing was added to a jacket, in the Arabian Gulf, and a crack was simulated on it. Thus, it was proved that the first method, called Vibrodetection, is able to detect and to localize a crack on a steel jacket, before the structural integrity of the whole jacket is affected. Then, the second method, a global one, was applied to determine the dynamic parameters of the structure (modes of vibration). A new computer code is now being realised to deduce from the measured dynamic parameters, the resulting integrity of the damaged platform.

The equipment developed to apply both methods offshore is now ready for operational application as demonstrated during the Arabian Gulf test. Complementary tests of stability will be made in the North Sea during 1981 to specify the threshold of detection of cracks by Vibrodetection.

1979

5th Round Projects

Title :Development of a Procedure for the Exploration of Areas with Poor Reflections by the Combined Application of Different Geophysical Methods, Taking the North West German Basin as an example	Project N° : 01.15/79
Contractor : PREUSSAG AKTIENGESELLSCHAFT ERDOL UND ERDGAS	Telephone N° : (511)19321
Address : Postfach 4829 Arndststrasse 1 – D 3000 Hannover 1 Technical director (or person to contact for further information) : Dr.F.Neuweiler	Telex : 922851

Based on the existing results of the nonseismic method, the programme has been devised to develop a system by means of which a concept can be established for the further exploration of basins where few reflections are obtained. Particular importance is attached to a comparative interpretation of the results of various geophysical methods and to summarising them in order to build up a picture of the geological structure of the basin concerned. Modern data processing methods will be used to a large extent. These developments are based on the north west German basin.

When the gravimetric data originating from the Reich Survey are processed, the influences of the varying densities down to the Zechstein base will be eliminated by approximation in two dimensional sections. The incorporation of seismic data to calculate the gravity effects of the sedimentary rocks down to the Zechstein base has been largely completed. Gravimetric "stripping" has started.

Based on a synopsis of the gravimetric results and the aeromagnetic results, a "basin study" will be compiled to illustrate the history of development of the north west German basin. At the same time, the calculation of magnetic models for several magnetic anomalies has started.

The processing of a number of selected long time sections (recording of deep reflections) has been summarised in a report. As a result, attempts have been made to map the Conrad and Mohorovicic discontinuities and also a possible "basement horizon". The existence of deep faults has been shown to be probable at two locations.

Prakla-Seismos GmbH is still investigating the usefulness of re-analysing the refraction seismic data obtained by the German oil industry during the sixties.

Attempts to improve the seismic processing results in the Pre-Zechstein area have not yet been completed. Investigations will be completed by model calculations.

151

Title : Pulse 8/RHO 3 Circular Positioning System	Project N° : 01.18/79
Contractor : SEA SURVEYS LTD. Address : Kinsale Road Rathmacullig West Cork – Ireland Technical director (or person to contact for further information) : M.Kavanagh	Telephone N° : 021-962600 Telex : 28442

The first chain of Pulse/8 stations to be installed around Ireland's offshore areas was in 1977 at Mizen Head, Co. Cork. Angle, Pembrokeshire and Pointe de Corsen, France. This provided positioning cover over the Celtic, Fastnet and Western approaches area. An extension of this cover to replace the inadequate temporary positioning systems previously employed in the Porcupine Bank area was made in early 1978 with the deployment of two additional stations at Hag Head, Co. Clare, and Aughleam, Belmullet, Co. Mayo. This gave hyperbolic position fixing cover over the Porcupine Bank area. However, due to the physical shape of the coastline, it is very difficult to avoid an unfavourable angle of cut of the positioning lines at the points of intersection over the Porcupine Bank and the expansion of the Pulse/8 patterns or grid caused by the distance from the base line. A solution to the problem would be "circular" ranging by carrying an "on board" time reference in the form of a high stability oscillator and the use of a sophisticated software package and mini-computer to resolve ambiguities and enable a time pedestal to be accurately set. Thus, the input from three shore stations could be resolved into a "circular" position line fix (3-way fix), (instead of a two line fix in hyperbolic mode) the most desirable and elegant positioning solution possible in this area. The quality of the fix would be excellent due to the wider angles of cut and lack of pattern expansion. This in turn would extend the possible area of coverage.

The equipment was designed and bench tested in 1979 and a field trial was carried out in 1980, although the tests proved successful further development of the system, incorporation increased computer memory, is in progress.

Title : Development Programme for Composite Offshore Jacket Structures	Project N° : 03.71/79
Contractor : WIMPEY LABORATORIES LTD.	Telephone N° : (01)573.7744
Address : Beaconsfield Road Hayes — Middlesex UB4 OLS Technical director (or person to contact for further information) : C.Billington	Telex : 935797

This project concerns the development of composite offshore structures consisting of piled steel tubular jackets with certain members either partially or fully filled with a cementitious material.

The principal design and construction problems associated with tubular steel structures concern the nodal joints, the design of which is controlled by static punching shear strength and fatigue considerations. These design problems are exacerbated in deeper waters where both static and environmental loadings are greater and considerable welding, post weld heat treatment and inspection problems result. The installation of steel platforms in deeper water is complicated by transportation limitations on the size and weight of structures.

A preliminary feasibility study carried out by Wimpey has shown that the use of steel and concrete composite construction could alleviate the joint design problem. In composite construction the materials are used efficiently in applications most suited to their mechanical properties. The cementitious material is contained and therefore greater strength and ductility is achieved. The steel tubular is the containment medium and therefore is predominantly subjected to hoop tension and the cementitious filling minimises any tendency for buckling of the steel shell and reduces the need for internal or external stiffening.

Test results on the static ultimate strength of composite tubular joints indicate that considerable economies might be possible. Work is continuing to establish empirical formulae to account for the effects of composite action on the joint strength.

An essential step in calculating fatigue endurance for tubular joints is the calculation of the stress concentration factor (SCF). Electrical resistance strain gauges were used to measure SCF's on test joints before and after grouting, prior to ultimate load testing. These results indicate significant reductions in SCF which imply large improvements in fatigue life or scope to economise on tubular wall thickness.

The long term fatigue endurance of grout filled joints is being investigated, pilot tests have been carried out and further testing is now in progress to study the fatigue performance with particular respect to the durability of the grout.

Title : Gravity Tower	Project N° : 03.72/79
Contractor : C.G. DORIS	Telephone N° : 584 11 64
Address : 58, Rue du Dessous des Berges F - 75013 Paris Technical director (or person to contact for further information) : D. Michel	Telex : 270263

The study has been split in two main phases:

- The first one is a feasibility study of an articulated structure to be installed in 350 m water depth, in North Sea conditions with a 30,000 head load. This part was completed in mid 1980.

- The second part is a very detailed study performed for a 490 m deep site, which was started in May 1980 and should be completed at the end of 1981.

In both cases, the studies included the following main items:

- Study of the hydrodynamic behaviour of the Gravity Tower, and assessment of the general bending modes.

- Optimization of the concrete floater shape and dimension, and of the steel lattice structure.

- Structural design of the above items and of the base and foundations.

- Development of the laminated articulated joint.

- Study of the drilling conductor guides submitted to bending due to the angular motions of the tower.

- Preparation of basin model tests to check the dynamic behaviour of the structure in place and during the main phases of construction and installation.

- Preparation of tests to be performed on the rubber articulation joint, to check that it can withstand fatigue.

— The studies performed in 1980 have confirmed the interest of the concept, since it has been possible to design a complete drilling production equipment with well heads on the deck, installed on a quite simple sub-structure, submitted to quite low motions. As an example, the following figures summarize the main results obtained in the detailed study.

Water depth	490 m
Typical site	Mediterranean
Head load	15,000 t
Extreme heel angle	2.9°
Concrete for the floater	31,000 m^3
Steel jacket	26,000 t

Title : Cryogenic Removal of CO_2 from Natural Gas	Project N° : 03.73/79
Contractor : SNAM PROGETTI	Telephone N° :02.53531
Address : C.P. 4169 20100 Milano Technical director (or person to contact for further information) : B. Cimino	Telex : 31246

The work has been started on Nov. 79 on the basis of the flow sheet established by Snamprogetti that proposes a cryogenic distillation process to remove high content of CO_2 from natural gas.

The project consists of a basic design, detailed engineering, procurement and construction of a prototype plant to treat a natural gas of Candela field at Ascoli Satriano in Foggia province (South Italy). The progress up to June 30th 81 was the completion of the basic design phase, and started the detailed engineering phase.

All the concerned technical services have been developed, the project and the equipment specifications are being issued one by one.

In accordance with the purchase requisition, the procurement department started looking for possible vendors asking them for supply of 4 or 5 pieces of equipment.

The technical quotations received from the vendors were submitted to concerned technical services who in turn, after a careful review of the offers, submitted tabulation sheets and technical appraisal for each equipment.

The refrigeration cycle package has become the most critical item with a delivery of 12 months. As a result the entire construction programme will depend upon this delivery period.

M/S AGIP have informed us that they are currently negotiating for the purchase of land at Ascoli Satriano, Snamprogetti will construct at their own expense the said experimental plant on the above mentioned land. The Snamprogetti's civil department, assisted by AQUATER Personnel has carried out on-the-spot investigation at the site, has established datum point coordinates and surveyed the land.

In addition soil investigations have been carried out to determine sub-soil geological formations thus soil bearing capacity.

Title : Concrete Articulated Tower Oil Production Platform "CONAT-OPP"	Project N° : 03.75/79
Contractor : BILFINGER + BERGER BAUAKTIENGESELLSCHAFT	Telephone N° :(040)22923—0
Address : Kanalstrasse 44 Postfach 760240 —2000 Hamburg 76 Technical director (or person to contact for further information) :	Telex : 211186

The CONAT Group designed in an extensive R&D programme an articulated monotower for offshore application and performed tests in the vicinity of the research platform "Nordsee".

On the basis of the know-how gained from these tests and the system components tested, a compliant platform system will be developed for drilling, production and eventually storage and loading of hydrocarbon products in water depths greater than 200 m.

The concept is based on a multi-column articulated system formed by an upper deck structure connected to the foundation by four legs with a universal joint at the ends of each leg, thus allowing the upper deck structure to move horizontally.

Extensive work has been done for hydrodynamic analysis of the structure's motion behaviour and the loads acting on the universal joints.

Water depth, number of legs, deck weight and geometrical shape have been varied in continuous consideration of the influence of construction phases with the object to find an optimum type of platform.

Inshore installation of required equipment is planned. Another focal point of work is the design of the underwater ball joints. Technical solutions of the piping penetration through the interior of the ball joint have been worked on. Due to the movement of the whole platform in seaway the piping through the ball joint must contain special flexible components which had to be devised. Different configurations of the ball joint piping system were designed; large-scale model tests of these important equipments were prepared, these tests should establish the components as trouble free.

Title : Systems for Recovery of Hydro- carbons from Small Offshore Fields	Project Nº : 03.78/79
Contractor : TAYLOR WOODROW CONSTRUCTION LTD. Address : 345 Ruislip Road Southall Middlesex UB1 22X Technical director (or person to contact for further information) : J. Smith	Telephone Nº :(01)578 2366 Telex : 24428

The objective of this project is to develop economic systems for the recovery of hydrocarbons from marginal fields, primarily in continental shelf water depths. The systems are being based largely upon articulating columns, adapting and extending technology developed throughout the earlier ARCOLPROD programme for deep water production (Contract Nº TH/03.21/76). The main new systems under development fall into three broad groups:

a) Free standing articulating columns to support nominal production, intermediate process or control functions on small isolated hydrocarbon deposits.

b) Columns performing riser and other ancillary functions for connection to floating production, primarily using semi-submersibles but also considering tanker options.

c) Columns for oil offloading to tankers which can surmount and house controls for subsea storage to improve export efficiency from marginal fields.

Establishing strong oil company collaboration has been an important aspect of the initial work, to ensure the acceptability of developed systems to the industry. Regular dialogue and correspondence is being maintained with Gulf, Marathon, Texaco, Phillips and Chevron, who have continued their collaboration from the original ARCOLPROD programme. For the new programme, BP and BNOC have joined the collaboration team, providing valuable experience from direct North Sea interests. All these companies have agreed to participate in a Steering Group for the project and a comprehensive review of the systems proposed has already been completed.

Work through 1980 has centred on the important background studies to the development. These included a start on establishing "design tool" analyses for motion response of columns coupled with floaters, backed by parametric hydraulic model studies in Taylor Woodrow's own wave tank facilities. Further test work on anchorage and tendon systems has also been substantially finalized to determine base design parameters. First order schematics of process layouts for a number of these systems have been drawn up to provide the design team with target space and payload requirements.

The study has already identified a wider range of options for the systems proposed, and the project is proceeding towards an interim review with the collaborators in the spring of 1981. This will be followed by a more intensive phase of development on preferred systems to a reference design and quality assurance standard, supported by comprehensive analysis and confirmatory testing of complete systems and components.

Title : Development of Design Techniques for Instability Problems in Offshore Structures	Project N° : 03.79/79
Contractor : **TAYLOR WOODROW CONSTRUCTION LTD.**	Telephone N° :(01)578.2366
Address : 345 Ruislip Road Southall — Middlesex UB1 22X	Telex : 24428
Technical director (or person to contact for further information) : J. Smith	

The objective of the research programme is to develop an efficient method for the accurate prediction of structural response and collapse load of concrete cylindrical shells under depp water hydrostatic loading. The results of the programme will be used to develop safe and economic design methods, thereby enhancing the potential of concrete structures for the exploitation and storage of offshore hydrocarbons.

The programme started in early 1980 with a comprehensive literature survey on the structural response and strength of hollow concrete cylinders under external pressure. This survey has so far identified the areas that require particular attention and the parameters that are likely to signif- icantly influence the strength of concrete shells under external pressure. Based on the available test data an empirical design method has been proposed and shown to provide a lower bound to almost all the test results available.

Phase I of the programme will continue by analytically investigating the effects, on the collapse load, of each of the parameters that have been identified from the literature survey. This work will lead to finaliz- ation of the scope of the large scale experimental model programme which is to be executed to augment the limited work already carried out in this field.

Preparatory work on test facilities selection and evaluation of instrumentation under high water pressures is well advanced.

Title : Testing of Production Techniques	Project N° : 03.80/79
Contractor : GERTH	Telephone N° : 749.02.14
Address : 4, Av. de Bois Préau 92502 Rueil Malmaison — Paris Technical director (or person to contact for further information) : M. Leblond	Telex : 69066F

Transfer of fluids from the sea bottom to the surface is one of the main difficulties that arise with the various systems envisaged for exploiting deep water offshore fields.

This project, which benefits from the results of earlier work, concerns sea trials of a riser foot manifold and experiments on a 6" pipeline to evaluate the behaviour of links for transferring diphasic fluids (gas and liquid).

Riser foot manifold

The riser foot manifold is a subsea set of valves distributing both the production of hydrocarbons and of fluids required for production (injection gas or water, recirculation or safety fluids). This equipment consists of modular elements that are reraised to the surface for maintenance. A test at sea in shallow depth was prepared to test the automatic positioning and correct connection of a manifold module on a structure simulating the base of a riser foot manifold. The test took place in October 1980 in Perthuis d'Antioche. It was interrupted following breakage of the drilling string used to position the equipment. Since this incident called neither the technique nor the method into question, the test was abandoned at this point so as to have available equipment for the deep water test planned for the Spring of 1981.

Diphasic flows

Experiments using a 6" diameter pipeline on diphasic flows covered the influence of the gradient (−5 to + 7°), pressure (15 to 40 bars), temperature and viscosity. Several mathematical models were preferred predict the pressure losses encountered when conveying fluids under diphasic conditions. The results are now being interpreted; the interest of this work resides in particular in the possibility of considering long distance conveyance through a single pipeline of the oil and the associated gas produced by marginal deposits.

Title : Subsurface Links for Deep—Sea Offshore Production	Project N° : 03.82/79
Contractor : GERTH	Telephone N° :749.02.14
Address : 4, Av. de Bois Préau 92502 Rueil Malmaison — Paris	Telex : 69066F
Technical director (or person to contact for further information) : M. Leblond	

Links between surface supports such as production platforms, loading stations or flare stacks can no longer use rigid tubes in increasing depths of water. The most promising solution would appear to be a link consisting of layers of flexible pipes submerged and suspended at their ends from the surface support to be connected, all the more so since this type of link would appear to be the only one that can feasibly be built in the high pressure operating range envisaged (300 bars).

This project consists in testing a flexible pipe hung by one of its ends from a surface support, the movements of which simulate those of a semisubmersible or ship, and connected at the other end to a buoy submerged at a depth of 60 metres.

The preparatory studies for this test have been completed, though the test itself has been postponed, pending the results of fatigue tests at the maximum service pressure of the flexible line to be used for this sea test. These fatigue tests, carried out ashore and not forming part of the present contract, are now nearing completion. The results obtained so far justify considering installing equipment for the sea test planned for 1982 at the FRIGG site in the North Sea.

A request for tenders to supply and install the various items of equipment is now in progress.

Title : Mobile, Floatable Offshore Platform	Project N° : 03.83/79
Contractor : DEUTSCHE BABCOCK ANLAGEN AG Address : Postfach 100347-48 Duisburger Str. 375 D- 4200 Oberhausen Technical director (or person to contact for further information) : E. Bitterlich	Telephone N° :(0208)8331 Telex : 0856 951

The project aim is a mobile, floatable Platform which being fixed to the seabed when operating can withstand the impact of ice under arctic conditions.

It was found that this could be achieved in a favourably economical way by applying ground-freezing technology to the purpose.

The first stage of the project therefore concerned the compilation, analysis and evaluation of existing design proposals. This first stage included the evaluation of relevant data and observations about arctic environmental conditions and the characteristics of ice and ice forces.

It was found expedient to provide the capability to resist ice impact separate from the normal platform features as oil production capability and personnel accomodation capability.

Therefore a concept is now followed which provides an artificial, frozen and resistant-to-ice-impact sand berm in the shelter of which floatable modules can be installed.

The further work concentrates on how to design and how to establish the relevant body of frozen material within the sand berm.

Title : Pilot Plant to Enhance Heavy-Oil Recovery. Ponte Dirillo Field, Italy	Project N° : 05.14/79
Contractor : AGIP S.p.A.	Telephone N° : 53531
Address : 20097 S. Donato Milanese 20100 Milano Technical director (or person to contact for further information) : M. Ciarichi	Telex : 31246

The purpose of this project is to design, build and put into operation a field pilot for an enhanced oil recovery process by non-miscible high-pressure gas injection into the very heavy (11° to 17° API) oil field of Ponte Dirillo, Italy.

The project is based on the results of a previous research, carried out by Agip S.p.A. and partially financed by EEC (Contract N° 16/75), on the heavy-oil field of Gela, Italy. This research project brought to evidence the fact that high-pressure gas injection (either natural gas or carbon dioxide) may considerably improve oil recovery from reservoirs of this type, through reduction of reservoir oil viscosity and swelling of reservoir oil.

The Ponte Dirillo field is a separate appendix to the Gela field, showing same reservoir rock and oil characteristics as Gela field.

A campaign of well surveys has been run on all Ponte Dirillo wells, to up-date production profiles throughout the pay zone. Production and TDT logs have been run.

The reservoir geology and engineering study has been updated, in view of a numerical model study of field behaviour. New PVT studies have been run on bottom-hole oil samples taken in three wells. The thermodynamic behaviour of reservoir oil saturated at various pressures with injection gas has been determined.

The plans for the surface facilities of the pilot have been completed. They include:

- a first stage compression unit of the associated gas from Gela field,
- a gas dehydration plant,
- a 4" pipeline from Gela to Ponte Dirillo (8 km),
- a second stage compressor unit in the Ponte Dirillo tank farm,
- a 2" high-pressure pipeline from the Ponte Dirillo tank farm to the injection well,
- a gas-oil separation and testing train, with related auxiliary tanks.

Title : Seismic and Geoelectric Frac Location from the Surface	Project N° : 05.15/79
Contractor : PRAKLA-SEISMOS GmbH	Telephone N° :(0511)80721
Address : Postfach 4767 Haarstrasse 5 D - 3000 Hannover 1 Technical director (or person to contact for further information) :M. Schiemichen	Telex : 922847

In order to enhance oil and gas production in existing and new fields, the oil industry uses with increasing frequency hydraulic fracturing of tight formations. In order not to interfere with the production of existing wells and to optimize the lay-out of new fields, it is of the utmost importance to know the direction of the faults created by hydraulic fracturing. The objective of the project is the detection of the azimuthal direction of these artificial faults by geophysical methods, either seismic or geoelectric.

For the seismic survey method it is planned to set up near to the injection well a cross- or star-shaped seismic array with a great number of receiving stations, each station consiting of 24 or more geophones. By the break-up of the formation during the injection of fluid, minor seismic signals are created at and near the fault plane. If it is possible to locate and pilot the hypocentres of these micro-seismic events, the azimuthal direction of the fault plane will become obvious.

The major difficulty and the reason why this method has not been employed yet, is the exceptionally unfavourable signal to noise ratio. The seismic signals are small and the noise created by the engines and pumps near the injection well is extremely high and covers a wide frequency range which is partly identical with the frequency range of the seismic signals. In order to improve the signal to noise ratio and to identify a seismic signal within the noise, intensive efforts in data acquisition and processing are necessary.

During a massive hydraulic fracture a noise recording has been taken over a distance of 225 to 575 m from the pumps, using groups of horizontal and vertical geophones. The analysis has been completed and revealed the peak of the noise from 12 to 30 Hz, with higher noise amplitudes on the horizontal components compared to the vertical components. The wave length of the noise train was approximately 30 m and the velocity between 280 and 820 m/sec. With this information it will be possible to optimize the geophone pattern during the forthcoming experiments.

Extensive processing tests showed that it is possible to improve the suppression of pumping noise by spike-deconvolution. In order to prove this, synthetic frac-signals were mixed with pump noise and the signals retrieved with different operator lengths.

For a future experiment it is planned to use additional 3-component geophones in shallow drill holes for comparison and signal detection.

Geoelectric measurements have not been performed so far.

Title : Development of a Pipelaying Method in Very Deep Waters with Mechanical Connections	Project N° : 09.17/79
Contractor : TECNOMARE S.p.A. Address : S. Marco 2091 30124 Venezia Technical director (or person to contact for further information) : M.Rodighiero	Telephone N° : 708622 Telex : 410484

Purpose of the project, started on December 1st 1980, is the determination and the development of the technology to be followed to arrive at the use of mechanical connections in single-station joining for pipelaying (in J-curve) in very deep waters. The foregoing statements allow to define the main features of the system specification:

- The system must allow the pipelaying in very deep waters in a J-curve
- The pipeline is assembled by means of mechanical connections
- The system must be used on a suitable vessel. The vessel must have a low daily rate
- The system must allow the laying of a sealine with the same reliability (or even better) of a sealine laid with the conventional method
- The system must allow to reach a laying cost considerably lower than the laying cost of conventional systems.

The main specification data are:

- Water depth : 1,000 : 2,000 m
- Fluid : oil or gas
- Pressure range : 0 + 30 Mpa
- Temperature range : 2 : 90°C

The main uncertainties of the project depend on the solution of the following problems:

- Definition of a low cost mechanical joint, with pressure tightness characteristics to allow its use in pipelaying in very deep waters
- Definition of the equipment for the pipelaying in J-curve
- Definition of the test equipment for the assembled joint
- Definition of the characteristics of a low cost laying vessel suitable for pipelaying in very deep waters.

Title : Automatic Processing of Side Scan Sonar Records	Project N° : 09.18/79
Contractor : SESAM SA (FRANCE)	Telephone N° : 890.82.44
Address : 132 Av. de Villeneuve St.Georges F - 94600 Choisy Le Roi	Telex : 202268
Technical director (or person to contact for further information) : P.Carmagnol	

1. Objective

The objective of this project is to develop a new method of processing of the side scan sonar records to be able to completely use all data gathered on such records. The present methods are either selective (only main features are taken into account) or very expensive.

2. History

Tentative methods have already been tested by Sesam which induced our company to develop this project proposed to the Commission of the European Communities.

3. Hypothesis of works

The data which are displayed on the side scan sonar records cannot be properly located with respect to the support vessel i.e. with respect to actual system of coordinates, because severals factors of uncertainty prevent these data from being correctly positionned: accuracy of the surface positioning, horizontal offset, heading of fish and support vessel, speed variations, height of fish above sea bottom etc... But, the data displayed on the records are very characteristic and the "acoustic" picture of a given object (pipe, wreck, reef, etc..) can be easily plotted on another record by human interpreter. A methodology has been developed to be able to automatically spot and locate various data corresponding to the same object.

4. Processing

Taking into account the approach of the problem as stated here above, we have defined the processing problems and how to solve them including the display of the results of this automatic interpretation.

Hardware

After a technical survey of hard ware available in France and Europe we have selected 3 possibles types of equipment: TEKTRONICS, WILD, BENSON. But on July 1981, we have spotted a brand new equipment manufactured by TECHDATO which induced us to completely review our plans as this equipment (video camera which digitizes and displays colour pictures) seems to be exactly adapted to our problem.

Title : Deep—Sea Repairs	Project N° : 09.19/79
Contractor : GERTH Address : 4, Av. de Bois Préau 92502 Rueil Malmaison — Paris Technical director (or person to contact for further information) : M. Leblond	Telephone N° : 749.02.14 Telex : 69066F

Evacuation of the oil and in particular gas produced offshore calls for the use of a large diameter pipeline, the laying and repair of which are problems that are well under control down to depths in the range of 250 to 300 metres. At greater depths, laying operations can still be carried out from the surface, though the repairing techniques for such pipelines are still in the preliminary stage. Whether they are to be repaired by welding (the WELDAP automatic welding techniques) or by mechanical connectors, they nonetheless call for very careful preparation of the ends, which must be carried out automatically, since divers can no longer work for extended periods at such depths.

The purpose of this project is to build and subject to sea trials fully automatic prototype equipment for preparing the ends of a large diameter pipeline underwater. This preparation ranges from cutting away the damaged ends of the pipeline, removal of the concrete and anticorrosion covering of the end of the tube and preparation by brushing and machining the longitudinal welds.

The prototype equipment consists of lifting devices and positioning structures on the bottom, a worktable carrying in turn the modules carrying out each of the basic tasks and a remote—control and transport device from the bottom to the surface.

The engineering studies of this equipment have now been completed, and potential suppliers of the equipment consulted.

Title : Rock-Burying Protection of Underwater Pipelines Submarine Vehicle "VELPO"	Project N° : 09.21/79
Contractor : DORIS - OTP - CEA - CFD	Telephone N° : 5841164
Address : 58, rue de Dessous des Berges F 75013 Paris Technical director (or person to contact for further information) : M. Martin	Telex : 270263

This project was started early in 1978 with the first phase which consisted in the preliminary study and feasibility study and has now reached on 12.31.1980 the end of the second phase which is the detail engineering and the construction of the submarine vehicle "VELPO". It is expected that the project will be complete by the end of 1981 i.e. the full scale tests at sea.

The "VELPO" project is issued from the wish to perfect an economical, reliable and highly efficient system for the rip-rap protection of underwater pipelines in deep water (down to 984 ft). It consists of a self propelled submarine vessel named "VELPO" stradding the line to be protected, the propulsion along sea bed being assured by either four crawler tracks or four Archimedean screw propellers, depending on soil conditions. The Velpo vehicle is equipped with a hopper guiding the rocks exactly above the pipe line. Rip-Rap to cover the line is dumped from a dynamic positioned surface vessel via a 24" vertical casing which funnels the gravel down and over the pipe line. The whole system (surface vessel and submarine vehicle) being totally uncoupled.

The submarine vehicle "VELPO" is equipped with highly improved acoustic sensors for locating the submerged pipe line and ensuring an accurate positioning above the line. Detection sensors are also arranged on vehicle to ensure permanent control of rip-rap bank shape above the pipe. The summarized operating criteria and main characteristics of the system are: for the waves (H = 6 m, period 8"), 3 knots current at the surface and 1 knot near the sea bottom, operating depth of 164 to 984 ft, 10% average gradient and 4% cross slope for the sea floor profil, 8" to 48" for the diameter of pipe-line to be protected with a bank section of 3.5 to 10.5 cu.m/linear meter. The dumping speed is 50 to 72 m/h and the dumping cubic capacity is 500 cu.m/hr.

The detail studies and reduced scale investigations and tests were carried out: on the shape and characteristics of the rip-rap bank, on the disphasic hydraulic flow of the gravels in the vertical pipe, on the behaviour of the rip-rap bank under current and wave action, on the mechanical study of the vertical casing with the displacement of its lower end and on the efficiency of the detection systems.

Advantages of the whole system are numerous: conveying of protection materials from a reference point located above the pipe line gives a better accuracy of rock dumping, shock absorption against the hopper walls avoid fragilisation of the pipe line, detection and acoustic sensors ensure an accurate positioning, adaptation to great depths in severe weather conditions and application pipelines already in service, permanent control with visualization on a screen.

Title : Deep Water Pipeline Repair	Project N° : 10.20/79

Contractor : SNAM S.p.A.
 	
Address : CP 3757
      20097 S. Donato Milanese
      20100 Milano
Technical director (or person to contact
for further information) : M. Bonfiglioli
	Telephone N° : 53531
	
	Telex : 31246

With the aim of providing efficient maintenance to the gas transmission lines between Tunisia and Sicily and those crossing the Straits of Messina, SNAM together with Snamprogetti, Saipem and Nuovo Pignone commenced a study of an automatic underwater repair station for deep water pipelines.

These studies should be completed in 1983.

This study is necessary because of gaps in the present repair technology. We have carried out an enquiry to identify those developed technologies which could be utilised in our development. The result of this enquiry was that no technology has been sufficiently proven at these depths.

We have completed a feasibility study which has enabled us to specify the principle requirement of the station and the type of connection necessary to carry out a repair. These specifications have been developed in order to optimise the operation of the station while Nuovo Pignone is developing the connection system.

The connection system between the pipe and the sleeve, the air tightness system between the sleeve and the spoolpiece have been built to full scale and successfully tested.

Scale models of the sleeve-spool piece junction and the flexible joint have been constructed and tests have demonstrated the feasibility of the concept.

The flexible joint may prove a viable alternative to spherical system presently on the market.

A full scale model of the spool piece is presently under construction.

Title : Repair of Subsea Pipelines by Mechanical Coupling	Project N° : 10.21/79
Contractor : GERTH	Telephone N° : 749.02.14
Address : 4, Av. de Bois Préau 92502 Rueil Malmaison — Paris Technical director (or person to contact for further information) : M. Leblond	Telex : 69066F

Subsea pipelines can be repaired routinely down to a depth of 200 metres of water, by welding in a pressurized atmosphere, known as the hyperbaric technique. Welding at atmospheric pressure, known as WELDAP, is itself limited by the depth of water owing to the dimensions and weights of the welding chambers. Fully automatic connection of these pipelines would mean that the repairs are no longer restricted by this limitation.

The purpose of the project is the construction and testing at sea of prototype equipment for fully-automatic subsea connection of a large diameter pipeline.

The principle retained is that of deforming the end of the pipeline into a collet into which a new section of pipeline can be inserted, by providing a metal-to-metal seal capable of withstanding the very high pressures and temperatures that are generally encountered.

— The formation of collets on standard pipeline tubes has enabled a method of forming by punching to be defined with no loss of the mechanical properties of the metal.

— Tightness tests enabled the nature and shape of the seals to be defined.

— A preliminary study enabled the devices for connecting to the end of the tube to be selected.

— Lastly, study of a fully automatic forming device has enabled the guide drawings of this device to be determined.

This work will continue in 1981 with tests of a reduced-scale connecting device (diameter 300 mm) and by testing of the components of the forming machine, prior to building a full-scale prototype.

Title : Development of a Deep Water Mooring System to Permanently Moor a Tanker in 400 m of Water	Project N° : 14.07/79
Contractor : SBM (UK) LIMITED	Telephone N° : (01)8913434
Address : Northumberland House 2a, King Street - Twickenham Middlesex TW1 3SN Technical director (or person to contact for further information) : R. Dyer	Telex : 28306

As it is now becoming common practice to use permanently moored floating production/storage facilities, this new project is aimed at exploring the next step in this technology. The maximum water depth for the existing systems available for permanently mooring the vessel is 150 metres. The water depth chosen for this next step was 400 metres under severe environmental conditions typical of the North Atlantic. Additionally, and in order to widen the scope of future applications, the effects of a milder environment were also considered at the same water depth.

The project was carried out in six successive phases, using theoretical studies combined with model tests to demonstrate the feasibility of several systems. Installation techniques were also studied in sufficient detail to show the feasibility of installing all the systems. However, the final choice of design will depend on site and reservoir conditions, number and type of risers, and client specifications, as well as the severity of survival conditions.

The six phases were: initial consideration of the mooring environments, assessment of existing technology and system selection, model testing, consideration of flowlines and risers, and methods of installation; the penultimate phase consisted of a system comparison. Finally, for one system a more detailed technical and economic analysis was performed.

The results of this project, including the two comprehensive model test series, were that engineering of the concepts developed could take place by scaling up proven components without the introduction of any novel technology. For milder conditions conventional systems could be designed and engineered without any significant scaling up problems.

1980

6th Round Projects

Title : The PSV Technique and Longitudinal and Transverse Wave Surface Seismics on the Deposit	Project N° : 01.21/80
Contractor : GERTH	Telephone N° : 749.02.14
Address : 4, Av. de Bois Préau 92502 Rueil Malmaison — Paris Technical director (or person to contact for further information) : M. Leblond	Telex : 203050

Study of transverse waves associated to longitudinal waves provides additional information on the lithology and fluids of hydrocarbon reservoirs. However, this surface method must itself be calibrated by making vertical seismic profiles for these two types of wave inside wells.

This project consists in performing this calibration on a gas deposit model during exploitation when the bubble of gas is at its maximum volume (Autumn) and minimum volume (Spring). Among other things, it must enable the procedures enabling the movement of the gas/oil/water interfaces during exploitation of a hydrocarbon deposit to be developed.

In 1980, an initial series of seismic profiles was made with longitudinal and trasverse waves. The quality of the results is satisfactory for the longitudinal waves. Modifications will be made to the procedure with a view to performing a second series of seismic profiles in 1981.

Title : High Resolution Seismics on Deposits	Project N° : 01.23/80
Contractor : GERTH	Telephone N° : 749.02.14
Address : 4, Av. de Bois Préau 92502 Rueil Malmaison — Paris Technical director (or person to contact for further information) : M. Leblond	Telex : 203050

The objective in high resolution seismics is to apply an economic method of seismic prospection enabling the seismic sections of interfaces separated by 5 to 10 metres to be distinguished.

The intensive pace of the search for new petroleum reserves indeed requires that a seismic tool capable of higher performance be available. Furthermore, the development of enhanced recovery techniques on fields that are already being exploited implies that finer geometrical definition of these fields be obtained so as to optimize the location of new wells.

A seismic campaign is to be made on the Marienbronn field in Alsace where an enhanced recovery pilot project is also to be set up.

In 1980, preparation for the campaign was completed. Exploitation of 3 seismic profiles was adopted, following experimental noise shots.

Provided the necessary administrative permission is obtained, the seismic campaign proper should be carried out in mid 1981.

Title : Development of Seismic Technology for Prospecting Hydrocarbons in Ante-Permian Coal Basins	Project N° : 01.24/80
Contractor : GERTH	Telephone N° : 749.02.14
Address : 4, Av. de Bois Préau 92502 Rueil Malmaison — Paris Technical director (or person to contact for further information) : M. Leblond	Telex : 203050

The objective of this project is to explore the deep-lying gas in the coal basins of Western Europe. Exploration of these basins is rendered difficult because of the depth of the products sought and the complexity of the geological structure of basins such as those of Namur and Dinant.

The purpose of the project is to define the best-suited seismic technology for exploring these basins, by tests made in situ and at the processing centre.

In 1980, 137 km of vibro-seismic profiles were recorded and processed. These 4 profiles were set out uniformly from West to East perpendicular to the Midi fault in the ante-permian basins of the Nord and Pas-de-Calais Departments of France. From the standpoint of implementation, the parameters required for satisfactory execution have been discovered thanks to this project.

The data acquired is now being analysed and interpreted.

Title : Development of a Steel Gravity Platform for 350 metre water Depth	Project N° : 03.74/80
Contractor : TECNOMARE S.p.A. Address : S. Marco 2091 30124 Venezia — Italy Technical director (or person to contact for further information) : M. Lucchetti	Telephone N° : 708622 Telex : 410484

The main purpose of the project, started on January 1st 1981, include:

- The identification of a fixed platform configuration and general structural arrangement and of a related installation procedure, in an average water depth of 350 m.

- The identification of the computer programmes and calculation procedures necessary to solve all the relevant naval and structural static and dynamic problems.

- The parametrical analysis of the various design specification items of the platform configuration and costs, in order to check the most convenient application range.

- The definition of the construction, transportation and installation techniques and procedures, related to well defined facilities.

- The development of the envisaged configuration(s) to an advanced engineering stage, in order to solve preliminary all the problems related to a possible industrial realization.

Title : Industrial Realization of a High Accuracy Counting System for Gaseous and Liquid Hydrocarbons	Project N° : 03.76/80
Contractor : ULTRAFLUX	Telephone N° : 979.26.40
Address : 63, Rue du Général de Gaulle F – 78300 Poissy	Telex : 696028
Technical director (or person to contact for further information) : M.M.J.Pierrat	

Stage 1 "Study of the try-out of the electronic device for automatic resetting of the basic electronics" has been completed. Stage 2 "Experimental study on the TRAPIL test loop for liquid hydrocarbons" gave rise to two series of experiments.

Preliminary tests have been carried out with a 6" three chord spool on the TRAPIL endurance loop. The purpose of the tests was mainly to determine the repeatability of the measurement by comparison with the indications of a reference turbine.

More than 250 measurements were effected on 2 different products (JP and FOD) between 0.5 and 7 m/s.

For flows having Reynolds numbers higher than 10^5 the repeatability is stabilized at 0.4 °/o of the measurement. The tests on the calibration loop itself with 3 chords of 16" have been effected. They have been analysed and shaped. The results are very good and enable to expect an accuracy quite consistent with the present requirements of the metering of liquids.

It has been decided to continue the tests, not on the TRAPIL bench but in the VERNON counting station; the test conditions have been set up with TRAPIL.

The mounting of the spool in line with the existing counting installation in VERNON will supply valuable data. The mounting is to take place in summer and the first results will be known at the end of 1981.

The study and construction of the 30" gas spool for the BERNOUILLI Laboratory of GASUNIE have been achieved.

The test programme has been defined with GASUNIE. It is planned to take place at the end of October 1981.

Title : Production of Methanol Offshore	Project N° : 03.84/80
Contractor : STONE & WEBSTER ENGINEERING LTD. Address : Stone & Webster House 236, Grey's Inn Road London WC1X 8HA Technical director (or person to contact for further information) : D.Bernard	Telephone N° :(01)857 2855 Telex : 299801

This project follows on from work undertaken by Stone & Webster Engineering Ltd's parent company on the development of an innovative reformer for nuclear applications.

This project is in part to develop a methane reformer as a pressurized high temperature enclosed heat exchanger, which through its lightness, compactness and safety features will have special application to an offshore environment. In full to develop the total system for an optimum location with the topside facilities built around the reformer and having a completely compatible floating support structure, storage and offloading system, so as to form a totally integrated unit that can be used to produce methanol from gas offshore.

Through protracted contract negotiations and delays encountered in the approval of funding, little progress was made during 1980, with agreement signature postponed until June 1981. However in March 1981 an effective start was made on the work and considerable progress has been made since.

Regrettably overall, six months progress has been lost on the commencement date of 1 October 1980.

Title : Self Installing Satellite Production Unit	Project N° : 03.90/80
Contractor : ATELIERS ET CHANTIERS DE BRETAGNE Address : Boulevard Prairie au Duc Cedex 2 - F 44040 Nantes Technical director (or person to contact for further information) : C. DEVAULX	Telephone N° : 473132 Telex : 710960

The design studies on the satellite production units has been aimed at the estimation of the general layout, and at the problems raised by anchoring such a unit with tensioned cables.

- The general layout studies have resulted in a total weight of 1130 tons, including 300 tons pay load.

- The mooring by tensioned cables has been studied, taking into consideration the fatigue characteristics of the cable. The system retained for the project is a direct tensioning system through hydraulic cylinders and accumulators.

- At the present stage of the project, the largest permissable wave is of 12 m. Beyond this value the dimensions of the structure would no longer be consistent with the planned service conditions, and the stresses transmitted to the production platform would no longer be acceptable.

  Basin tests will be conducted at the ENSM model basin in Nantes. The model, at 15/1000 scale, is presently being fabricated.

Title : System for Subsea Oil Wellheads	Project N° : 03.91/80
Contractor : THE BRITISH PETROLEUM CO.	Telephone N° :01.920.8000
Address : Britannic House Moor Lane — London EC2Y 9BU Technical director (or person to contact for further information) : D.K. Knights	Telex : 888811

The project involves the development of a system for monitoring multiplexing and transmitting data, derived from a number of analogue and status sensors, mounted on a subsea production wellhead, where diver intervention is not possible.

The work is divided into the following phases:

— Phase 1 — Sensor development and testing.
— Phase 2 — Total system development and operational testing.

Work commenced March 1980 and competitive bids were obtained for Phase 1 of the project. A contract was awarded to TRW Ferranti Subsea Ltd. to cover development of:

a) Valve Status Sensor
b) Through Flow Line Train Sensor
c) Pressure Transducer
d) Temperature Transducer

The various sensors/transducers are presently undergoing qualification and long term test procedures which are anticipated to be completed by 1st July 1982. At this time the detailed requirements of Phase 2 will be identified to allow the project to proceed.

In parallel with the main development Bradford University are conducting a survey of techniques, which could be applicable to valve position sensing on wet subsea wellheads, which is expected to be completed by April 1982.

Title : Heavy Oil Pretreatment Platform	Project N° : 03.94/80
Contractor : GERTH	Telephone N° : 749.02.14
Address : 4, Av. de Bois Préau 92502 Rueil Malmaison — Paris Technical director (or person to contact for further information) : M. Leblond	Telex : 203050

The exploitation of immense reserves of non—conventional heavy oils raises the difficult problem of transporting them from the production fields to shore terminals or refineries. The only solution to this problem that can be envisaged is to pretreat the oil on the production field itself.

From the technical standpoint, two approaches can be considered to achieve this.

- The thermo—physical process, enabling the light fraction (deasphalting) to be separated from the heavy fraction used for production requirements (steam).

- The thermo—catalytic process, using pressurized hydrogen enabling the heavy oils to be cracked while at the same time reducing their viscosity.

In order quickly and efficiently to industrialize the methods employed for pretreating heavy oils at the field, an experimental platform with a capacity of 15,000 tons/year, will be built at the FEYZIN site.

In the initial stage, with which the present contract is concerned, the process documentation and basic engineering for the platform will be developed. At the same time, the search and tests on a laboratory pilot scheme will enable new technologies to be developed and methods so far applied in the refineries for conventional oils to be adapted to heavy oils.

The following work was completed in 1980:

- drafting of the process documentation and files, both for the main units (desalination, deasphalting distillation, visco—breaking and hydrotreatment) and the utilities.

- the basic engineering for the platform.

As regards the laboratory pilot tests, study has been started of the various methods of applying such processes to crudes that are at present available in large quantities (Boscan, in Venezuela, and Athabasca, in Canada).

Title : Development of Single Well Oil Production System	Project N° : 03.101/80
Contractor : THE BRITISH PETROLEUM CO. LTD.	Telephone N° :(01)628 4090
Address : Brittanic House Moor Lane — London EC24 9BU	Telex : 888811
Technical director (or person to contact for further information) : P.Haywood	

The project involves the development of an itinerant floating production and storage system based on a special purpose vessel provided with station-keeping equipment and a riser system for connection to the sea bed equipment.

The work is divided into the following Phases:

Phase 1. Feasibility study.
Phase 2. Detailed design and prototype testing.
Phase 3. Construction and operational assessment.

The present Contract with the EEC only provides support for Phases 1 and 2.

Work commenced on Phase 1 in December 1979 with the production of a detailed scope of work for the Feasibility Study. The study involved four sections:

Section I   Subsea Equipment
Section II   Production Riser
Section III   Process Plant
Section IV   Vessel

A contract was placed with Vickers Aker in January 1980 for the overall co-ordination of the Feasibility Study, the work of Sections II and III and collation of the final report. Sections I and IV were produced in-house by BP together with a system analysis and an economic study. The conclusion was that a new building was economically preferable to a conversion.

In addition to the foregoing two alternative studies were carried out; a turret moored production tanker by The Offshore Company, Houston, and a single anchor leg moored tanker by BP in-house. Both alternatives were rejected in favour of a new building.

Following completion of the Feasibility Study in April 1980 and in parallel with subsequent Phase 1 investigations a scope of work was prepared for the detailed design study for a facility based on a new tanker building.

Competitive bids were obtained for the detailed design study of the vessel, for the management of the work and for detailed design study of

the riser and process equipment. However, no contract was awarded at this
time as in view of the long engineering and fabrication programmes associated with a new building, in September 1980 BP initiated an investigation
into the conversion of a smaller existing vessel, "British Loyalty".
The resulting cost estimate prompted a reassessment of the new building
costs particularly to take into account pertinent legislation.

Title : Enhanced Oil Recovery from Egmanton Oil Field by Carbon Dioxide Miscible Flooding	Project N° : 05.15/80
Contractor : THE BRITISH PETROLEUM CO.	Telephone N° : 01.9208000
Address : Britannic House Moor Lane — London EC2Y 9BU	Telex : 888811
Technical director (or person to contact for further information) : D.L. Knights	

The project is concerned with the injection of carbon dioxide into the central well of an inverted five-spot pattern in order to miscibly displace part of the residual oil contained within the pattern. The work is at present divided into:

Phase I   Feasibility Studies
Phase II   Pilot and further studies
Phase III   Application of the results of Phases I and II

The present contract with the EEC only provides support for Phases I and II. BP have entered into an agreement with la Compagnie Française des Pétroles, la Société Nationale Elf-Aquitaine (Production) and l'Institut Français du Pétrole whereby these French companies (collectively known as the GERTH partners) will perform part of the studies. Project expenditure (less the appropriate EEC grant) on the studies section of the project will be met: 50% by BP and 50% by the GERTH partners.

Phase I Studies have been progressing well and a number of aspects covered. These include: PVT analysis of Egmanton crude and interfacial tension measurements, routine and special core analysis on core samples from well EG 68, and displacement studies through packed sand column and through composite core samples (again using core from EG 68). In addition IFP have carried out a review of the available reservoir data.

Activities on the Phase II pilot have been somewhat delayed. An initial design for surface facilities was found to be too costly (at £2.9 million). However work has gone ahead on producing a detailed design for a revised surface facilities concept of estimated cost £1.4 million. Savings were achieved principally by eliminating the test separator facilities and by simplification of the data recording system.

Well Egmanton 68 has been drilled (as the central well in the five spot pilot area) and completed as a producer. In addition a programme of workover and stimulation work on producing wells in the pilot has been initiated.

Title : Down Hole Steam Generator	Project N° : 05.16/80
Contractor : B.P. Research Station	Telephone N° :
Address : Sunbury-on-Thames	Telex :
Technical director (or person to contact for further information) : **Dr. T. Lister**	

1. This report describes the current state of development of a down hole steam generator using a pulsed burner under EEC contract TH/05.16/80.

2. A computer controlled rig has been designed and built to enable the testing of pulsed burners and ignition systems at pressures up to 70 bar. Instrumentation problems have, however, delayed the full commissioning of the rig, but burners have been tested at pressures up to 10 bar using the manual operating mode.

3. A number of high energy ignition systems have been examined and developed. These are:

   a) A tri Ethyl Aluminium ignition system has been developed by Schering AG within this contract. This was found to give unreliable ignition when tested within an atmospheric pulsed burner.

   b) A plasma plug has been developed at Sunbury in collaboration with Imperial College, London. This system gives reliable ignition of gases by emitting a high energy plasma jet. The plasma plug reduces electrode erosion and breakdown voltage requirements which limits the operation of conventional high energy spark plugs.

   c) A hydrogen flame tube ignition system has also been developed within the contract. This has the advantage of giving a high energy ignition source from a very low energy spark discharge.

4. Preliminary preparations for a UK field trial have been undertaken which include:

   a) Selection and completion of a suitable well in the Egmanton Field.

   b) Injectivity testing on the selected well to establish rock permeabilities.

   c) Detailed cost analysis for the site preparation and field trial equipment.

191

Title : Steam Injection into the Marienbronn Heavy Oil Reservoir	Project N° : 05.19/80
Contractor : GERTH	Telephone N° :749.02.14
Address : 4, Av. de Bois Préau 92502 Rueil Malmaison - Paris Technical director (or person to contact for further information) :M. Leblond	Telex : 203050

The aim in implementing this pilot scheme to inject steam into the Marienbronn deposit is rapidly to acquire the knowhow that will be indispensable if the cpacity of oil companies to intervene in countries, particularly in Europe, containing immense reserve of heavy oils is to be increased.

In the initial stage of this project, which is the purpose of the present contract, the very high viscosity of the Marienbronn oil requires local stimulation of each individual well in the field, by a cyclic steam injection method (HUFF and PUFF).

If heating occurs homogeneously, consideration can be given to continuing the pilot experiment within the framework of a new project by applying an inter-well steam drive method.

The preliminary studies were completed in 1980. At the same time, various negociations were launched with a view to obtaining all the administrative authorizations and permits needed to implement the pilot scheme (impact, safety, soil occupation, etc... studies).

Title : Industrial Pilot Project for Injecting Microemulsion and Polymers into Neocomian Deposits	Project N° : 05.21/80
Contractor : GERTH	Telephone N° : 749.02.14
Address : 4, Av. de Bois Préau 92502 Rueil Malmaison — Paris Technical director (or person to contact for further information) : M. Leblond	Telex : 203050

Among methods of enhanced recovery of oil, the injection of micro-emulsions and polymers is the most attractive technique, enabling up to 70% of the oil in place to be recovered in certain cases.

This method has already been studied with the financial assistance of the EEC within the framework of a pilot method set up at CHATEAURENARD on a Neocomian formation (Paris Basin).

The implementation of an industrial pilot project, about 50 times greater in size than the pilot method is an unavoidably necessary step before taking the highly important decision as to whether to extend the method up to the scale of the formations of the Neocomian.

In the initial stage, which is the subject of the present contract, laboratory studies will enable the nature and optional volume of chemicals required to be defined. At the same time, studies of the deposit will enable the indispensable boreholes to be drilled. Likewise, the engineering design of the units for injecting the chemicals will be carried out so that the microemulsion injection and surface installations can be built at a later stage.

In 1980, difficulties were encountered at laboratory level in finaliz-ing the composition of the microemulsion. However, four injection wells and one production well were drilled and the basic engineering of the surface installations defined.

Title : Underwater Pile Driving Test of an Offshore Pile Driver Hydraulically Operated by an Underwater Powerpack	Project Nº : 06.08/80
Contractor :KOEHRING GmbH – MENCK DIVISION	Telephone Nº :(04106)6141
Address : Postfach 40 D – 2086 Ellerau	Telex : 02 13294
Technical director (or person to contact for further information) :Dr. J. Broer	

The complete pile driver plant, comprising the electrical system, winch for power and control cable umbilical, underwater powerpack and pile driver, has been available for testing purposes in April 1981. In order to ensure faultless operation of the pile driver rig the testing procedure was grouped into the three stages which are outlined below:

I.   Operational test with underwater powerpack ashore at the MENCK plant in Ellerau.

II. Submerging test with underwater powerpack in the Seine river at the location of the winch manufacturer, and submerging test of the umbilical at the COFLEXIP company in Le Trait.

III. Underwater pile driving test with the pile driver, incorporating the underwater powerpack, in the SHELL/ESSO offshore field "North Cormorant".

For the purpose of performing the operational test, as stated under I. the powerpack was connected to the local power supply; via electrical power system of the MENCK plant. During the course of the test one electric motor powered pump unit at a time was started and allowed to operate against the safety valve of the hydraulic system, that is, it was operated at maximum load. It was not possible to operate all electric motor powered pump units simultaneously; the mains did not supply the max. load of approx. 2,600 KVA, but only the 1,700 KVA starting power for one pump unit, the purpose of protecting the powerpack against impact stress. During the course of the equipment handling involved with the continuation of the test operation, the specially made umbilical, holding the power supply and control cables for the equipment, was unfortunately seriously damaged on account of inadequate preparation of the handling job and carelessness of the personnel aboard so that a continuation of the operation was rendered impossible, and thus the test was terminated on May 29. Following this, the pile driving plant was dismantled, stowed and then transported back to shore (activity "I") (Figures 14 and 15).

In general, the test result proved to be most favourable.

Title : Adaptation of a Gravity Foun- dation System to Removable Production Platforms	Project N° : 06.10/80

Contractor : SEA TANK CO.
 	
Address : Immeuble IENA
  12, rue le Corbusier
  F - 94588 Rungis Cedex
Technical director (or person to contact
for further information) : M.Letourneur
	Telephone N° : 687.2332
	
	Telex : 200939

Concrete gravity platforms have the advantage of being installed on site by its own means, due to its buoyancy and incorporated ballast system. Sea-bed stability is achieved by self-weight. It may be necessary to provide shallow skirts to prevent sliding.

Reversibility of ballast operations and the use of a raft foundation give these structures a distinct advantage concerning retrieval and reuse at another site. However, an uncertainty exists with respect to the nature and magnitude of bonds at the soil structure interface. A bibliographical study allows us to conclude that for a foundation consisting of granular material there is no adhesion problem. On the other hand, various experiments (in laboratory and in situ) have shown adhesion phenomena when the sea-bed is made of clay. The breakout forces needed to retrieve the structure are of the same order of magnitude of the bearing capacity of the soil. However, most of these experiments were performed with clay having an undrained shear strength much less than 10 kPa, and with objects that were generally deeply embedded in the soil.

A model test laboratory study on clay samples having an undrained shear strength greater than 50 kPa, under an hydrostatic pressure of 1 MPa, simulating the actual installation conditions of a platform should provide us with a better understanding of these phenomena. The proposed research includes

- selection of soil characteristics
- complete bibliographical study
- changes in soil characteristics under the loading of the platform
- experimental study of suction at the soil – structure interface ; means to accelerate its dissipation. Relationship between unsticking – tearing up and punching
- theoretical study of skirt breakout
- realization of the technologies needed for platform retrieval.

The bibliographical study has been completed and the laboratory tests are still in progress.

Title : Investigation into the Behaviour of Piles under Tensile Loads	Project N° : 06.12/80
Contractor : TAYLOR WOODROW CONSTRUCTION LTD. Address : Taywood House 345 Ruislip Road Southall, Middlesex UB1 2QX Technical director (or person to contact for further information) : Dr. D.Brunton	Telephone N° : 687.2332 Telex : 200939

The project is directed towards developing a procedure to reliably and economically design piles under the severe conditions of combined tensile and cyclic loadings. Present level of understanding is not adequate to predict tension pile behaviour under these loads, which would be imposed on anchorages in a number of proposed new applications.

Present effort is concentrated on general preparatory work and the detailed design of the field scale test programme. New published information on the response of soils under cyclic loading is being consolidated into this work.

Over the next few years, the programme will include field pile tests, laboratory tests on representative soil samples and the theoretical modelling of soil pile interaction. The influence of soil plasticity, particle size grading and consolidation history on pile behaviour will be assessed by covering a range of soils. To reduce scaling effects, the performance of 10 metre long piles will be monitored in the field tests; considerable instrumentation will be built into the piles to record response to both static and cyclic loads. The results will be compared to laboratory test results and the theoretical predictions. The final recommendations from this work should be in the form of a design method based on simple analytical methods, using soil parameters as measured during an offshore site investigation to determine input parameters.

Title : Remote Controlled Intervention System	Project N° : 07.35/80
Contractor : ATELIERS ET CHANTIERS DE BRETAGNE Address : Boulevard Prairie au Duc Cedex 2 — F 44040 Nantes Technical director (or person to contact for further information) : C. Devaulx	Telephone N° : 473132 Telex : 710960

A. Preliminary studies

1. Study of the underwater oil plant jobs

For each type of operation, e.g.: pipe laying or repair, a complete list of all the successive jobs has been drawn up.

2. Description of the tooling

The tooling specific to each operation is listed in tables.

3. Draft study of the architectures of the suitable systems
The following selection has been made:

- Configuration of the system: it comprises a slide ballast, and the craft can work in light or heavy mode.

- The power system consists of a coaxial umbilical current carrying cable, single-phase/medium voltage, and the video and teletransmission signals disconnected on the craft.

- An electro-hydraulic power plant on board the craft controls all the units. Motor power: 50 kw.

- The buoyancy of the craft is controlled by four regulators.

- The craft can be connected to various peripheral tools.

4. Definition of the main components and critical sub-systems

In process.

B. Study of the tooling

The following are defined:

- The power manipulator (6 degrees of freedom)

- The power plant is hydraulic (the electric motor is being designed)

C. System pilot and ergonomy study

In process.

D. Study of the operating means

Are defined:

Environment: We have studied the main movements of naval supports (Supply Boat Type) for sea forces ranging from beaufort 3 to 9. Furthermore, the "Slide Ballast" has been designed to operate in marine currents up to 3 knots.

The handling system with heave motion compensation which comprises a cart hauled by two winches. This cart is connected to the surface craft. The umbilical cable is stocked on the winch located on the ship bridge. The device studied enables to work up to force 6 and recover the craft up to force 7.

E. Study of the energy aspects

- Analysis of the mission profiles in the power—energy field. The jobs listed in the tables are estimated in power need.

- Analysis of the usable power sources.
  - The autonomous sources have been studied, and an energy schedule has been drawn up. They have a prohibitive weight for the underwater craft.

    In the present critical state, the only valuable source for one thousand metres depth is the surface generated electrical power and carried by the umbilical cable.

- The optimum energy system in exploitation cost is, in the present technical state, that mentioned in A-3.

Title : Personnel Transfer Unit	Project N° : 07.36/80
Contractor : **ATELIERS ET CHANTIERS DE BRETAGNE** Address : Boulevard Prairie au Duc Cedex 2 — F 44040 Nantes Technical director (or person to contact for further information) :C. Devaulx	Telephone N° :473132 Telex : 710960

The project objective is to built a transfer unit for personnel or light loads, with the following capabilities:

- lifting/lowering cab off/to ship deck at a specified point, with no impact or significant acceleration, irrespective of ship motion in the horizontal plane

- operating under severe environmental conditions

- improving transfer safety

The cab may be transferred from a fixed or semi-submersible platform, or from a barge equipped with a crane.

The unit has been built and is currently being tested in our plant. P.T.U. equipment and ancillaries have been tested. Some development problems have been encountered, which are now resolved, or being resolved.

The sea trials which are to follow will provide verification of P.T.U. performance under the operating conditions specified at the design phase.

A patent application has been filed in 1980, and is now pending.

Marketing actions in France and other countries have elicited a highly favorable response.

Title : J—Configuration Laying Electron Beam Welding	Project N° : 09.19/80
Contractor : GERTH	Telephone N° : 749.02.14
Address : 4, Av. de Bois Préau 92502 Rueil Malmaison — Paris Technical director (or person to contact for further information) : M. Leblond	Telex : 203050

The method of laying pipelines in a J—curve turns out to be highly interesting for depths of over 600 metres of water. However, the rational application of this method requires that a method of fast welding at a single welding station be developed.

The technological development envisaged consists in building and testing a prototype machine for electron beam welding of pipelines.

A complete prototype machine for welding 24 to 32" diameter pipelines has been built. Tests in the workshop have shown that the welding rate was over 4 butt—welds per hour, enabling pipeline laying rates that are competitive with the best known methods to be envisaged.

The construction of a laying ramp and a device for shore simulation of the operation as a whole (pipeline abutment and laying) in great depths of water is now being constructed so as to qualify the method industrially.

Title : Storage of Low Temperature Liquefied Gases in Clay	Project N° : 14.10/80
Contractor : DISTRIGAS S.A.	Telephone N° : 230.50.20
Address : Avenue des Arts, 31 B - 1040 Bruxelles	Telex : 23317
Technical director (or person to contact for further information) : M. de Lannoy	

The work has been divided into two distinct phases:

1) A study phase aimed at adapting the results of contract N° 17/75 to the present case

2) A second phase involving the construction of a pilot cavern based on the results of phase 1.

The conclusion drawn from the first phase was that the feasibility of storing LPG at -45° C in clay has been proven theoretically and that the chances of success at -160° C are sufficient to justify the extension of the cooling until this temperature is reached.

In the second phase SMET BORING have been engaged to dig out the test cavern by tunnelling. The access galery is supported by arches in reinforced concrete, while the supports of the test cavern are of two types:

- a special non reinforced concrete
- a cement based non reinforced concrete

The cooling will be accomplished by using azot liquid. During the cooling and test periods the cavern will be monitored by the following devices:

- 192 temperature measuring devices implanted in the ground and 12 measuring points in the cavity itself.
- Measurements of cavity displacement by vibrating cord.
- Measurements of groud displacement by extensiometers and inclinometers placed in boreholes.
- Measurement of micronoise.
- Location of the cold front by seismic refraction.
- Visual supervision by close circuit TV.

The current planning envisages the start of refridgeration in September 1981 and the conclusion of testing at the end of May 1982.

Title : Meteoceanographical-Structural Measurement System for the Safety of the Nilde Single Anchor Leg Storage - Sicily Channel	Project N° : 15.11/80
Contractor : AGIP S.p.A.	Telephone N° : 53531
Address : C.P.4174 - 20097 S. Donato Milanese 20100 Milano Technical director (or person to contact for further information) :M.Ciarichi	Telex : 31246

The main aim of the project is to verify the correspondence of the present design methodologies used for a "Single Point Mooring System" for NILDE oil field, by comparing the stress measured values, in the critical structural elements, in the different buoy-tanker configurations due to the environmental parameters (waves, wind, current), with the values obtained by a model test programme performed in a preliminary designing stage.

The SALS (Single Anchor Leg Storage) concept has been developed recently mainly for the exploitation of many secondary oil-fields. The project deals with a dynamic and structural data acquisition system. The system measures and records the meteoceanographic data, the dynamic parameters of the riser, yoke, tanker and the structural parameters in some critical elements of the SALS. The collected data, are then pre-elaborated and interpreted to reach the following target:

- Evaluation of the dependability of the SALS design methodologies
- Evaluation of the effect of the model scale
- The continuous check (during the exploitation) of the critical structural elements

To date 31.12.1980, all the equipment has been procured and it is ready to be installed either on board of the tanker or the yoke. Present forecast is to start with the field measurements toward the beginning of 1981.

Title : Development of Methods of Rehabilitating Damaged Offshore Concrete Structures	Project N° : 15.16/80
Contractor : TAYLOR WOODROW CONSTRUCTION LTD. Address : Taywood House 345, Ruislip Road Southall – Middlesex UB1 22X Technical director (or person to contact for further information) :M. Papworth	Telephone N° :(01)5754646 Telex : 24428

The objective of the research programme is to produce suitable methods of repairing major structural damage to concrete structures. This covers the three following specific areas, all of which will be considered during the programme:

1) Assessment of structural condition
2) Providing access to the repair area
3) Restoration of Active and Passive reinforcement and replacement of concrete

Computerized techniques are being investigated to assist the damaged state, possibly before a detailed inspection, resulting from ship and other impacts. Two independent programmes have been developed to determine the load deflection curves for a ship and a concrete caisson. This is basically represented by three non-linear springs to determine the load time relationship for the deformation of the caisson during a collision. As a result, the damage assessment can then be computed using a dynamic relaxation programme.

As repairs are likely to be relatively complex, methods of providing a suitable working environment at the repair surface are being investigated. These include atmospheric chambers for the splash and underwater zones and sealing mattresses which can be rapidly installed in emergency situations.

Restoration of the prestressing state, including analysis of the stress states in the local damage area, is necessary in order that the overall structural integrity, during extreme environmental conditions, is restored. This assessment is being carried out using quarter scale models which will be tested to represent the damage, repair and specific environmental loading situations. Methods of coupling reinforcements and replacing concrete are also being investigated.

Throughout the programme the practicality of the repair methods is being ensured. This will be proven in the final phase where underwater trials on large scale repair sections will be undertaken.